multivariable technical control systems

multivariable technical control systems

roceedings of the 2nd IFAC Symposium
uesseldorf, October 11-13, 1971

. Schwarz
ditor

urvey Papers-Volume 4

971
orth-Holland Publishing Company — Amsterdam · London
merican Elsevier Publishing Company, Inc. — New York

North-Holland ISBN for complete set: 0 7204 2055 5
North-Holland ISBN for this volume: 0 7204 2059 8

PUBLISHERS:

NORTH-HOLLAND PUBLISHING COMPANY – AMSTERDAM
NORTH-HOLLAND PUBLISHING COMPANY, LTD.–LONDON

SOLE DISTRIBUTORS FOR THE U.S.A. AND CANADA:

AMERICAN ELSEVIER PUBLISHING COMPANY, INC.
52 VANDERBILT AVENUE
NEW YORK, N.Y. 10 017

PRINTED IN GERMANY

PREFACE

Led by the experience gained from the 1st Symposium on Multivariable Control Systems, Duesseldorf 1968, the International Programme Committee of the 2nd Symposium tried to improve the efficiency of this meeting by the following:

- restriction to a relatively small number of topics seeming particularly relevant for technical multivariable control problems,
- improvement of the understanding of the relevant problems by inviting survey papers presented by internationally known experts.

The International Programme Committee, whose members are cited in the preface of the volumes 1-3, established three theoretical and two application topics:

- Linear Multivariable Systems Theory
- Optimal Multivariable Control Systems Theory
- Multilevel Technical Systems Theory
- Applications in Process Industries
- Applications in Power Plants and in Power Distribution Systems.

The particular reason to choose application topics cited above was that in these industries multivariable control problems are known for more than 30 years. Another reason was that only those topics should be treated which are not covered by other IFAC Symposia.

In order to publish the proceedings of the papers presented and discussed at the 2nd IFAC Symposium on Multivariable Technical Control Systems immediately after the conference the organizer, the publisher and the editor agreed to publish the unrevised papers for which the text of the preprints will be used. For technical reasons and in order to save time the technical papers are gathered up in the volumes 1-3 whereas volume 4 contains the five survey papers though the last were presented indeed at the beginning of each concerning session.

The authors of the invited survey papers had been asked to present their papers in such a way that it may serve as a survey and as a tutorial also for those participants who are not up-to-date with the most recent developments. The editor feels that all five authors succeeded extremely well in solving this challenging problem.

In the first contribution of this volume A.G.J.MACFARLANE gives first a rather detailed survey of those methods of linear multivariable control system theory being at hand today. A highlight of this paper is the presentation of generalisations of classical frequency-response-methods originally introduced by BODE and NYQUIST. This paper contains many results originally due to the author. A rather voluminous bibliography is an exciting source of references.

The second contribution presents various methods for solving technical optimization problems. M. THOMA points out some fundamental properties of optimal control and different optimization principles. His paper together with a selected reference list of 94 titles will be a good tutorial guide for all who are not yet confronted with optimization problems in the field of multivariable control.

Among the striking characteristics in the evolution of the practice of control system theory is the aspect of optimum control of large-scale and multilevel systems. A surprisingly small number of technical papers had been presented for the topic of Multilevel Technical Systems. The more important for this meeting is the third contribution in this volume by M.D. MESAROVIC who is the leading worker in this important field of multivariable control systems theory. His paper gives a nice idea of some fundamental properties of multilevel control systems. Especially he indicates some research opportunities which are of greatest immediate need for the solution of recent practical problems.

In the fourth contribution J.E. RIJNSDORP who is one of the leading workers in this field demonstrates how multivariable control theory may be applicated in process industries. The problems of Analysis of Interaction, Invariance Techniques, Modal and Pole-shifting Techniques, Optimal Control Techniques and Identification and Adaptation Techniques also are of particular interest. An extensive bibliography may in addition help to improve the understanding of control problems in process industries.

A comprehensive treatment of the application of multivariable control theory in power plants and power distribution systems is presented in the following fifth contribution by J.P. WAHA. He describes techniques necessary to solve the integrated problems involved with such complex integrated control systems. Among these techniques are Identification and Modelling Techniques, Techniques for Solving High Dimensional Problems, Optimal Control Techniques from an unilevel as well as from a multilevel viewpoint. This paper is a good cross-section of multivariable control theory and of complex power plant and power distribution systems as well.

Though each of these five authors gives his particular view of the field he is mostly interested in and though these fields seem to be quite different at a first glance, these five papers are really interrelated nicely by the aspect of multivariable control systems theory. This volume together with the technical papers in the volumes 1-3 demonstrates once again that multivariable control means more than a purely formal generalization of single variable control experience.

I wish to thank the authors and the publisher for their kind and understanding cooperation.

October 1971

H. Schwarz
Editor

Survey Papers - Volume 4

Contents

Contents

Technical Papers - Volume 1

Contents

Technical Papers - Volume 2

Contents

Technical Papers - Volume 3

A.G.J. MacFarlane

LINEAR MULTIVARIABLE FEEDBACK THEORY: A SURVEY

LINEAR MULTIVARIABLE FEEDBACK THEORY: A SURVEY

A.G.J. MacFarlane,
Professor of Control Engineering,
University of Manchester Institute of Science & Technology,
Manchester/England

CONTENTS

1. INTRODUCTION

Over the past few decades a vast amount of research has been expended on linear multivariable feedback theory and its application to the design of feedback control systems. In order to present a coherent and manageable survey of such an enormous body of literature, it is necessary to adopt a particular point of view and present everything in terms of it, even at the risk of over-emphasis of those specific aspects of the field which are most naturally treated in this way. The viewpoint adopted here is that the natural approach to feedback theory is via the frequency-response methods originally introduced by Bode (Bode, 1945) and Nyquist (Nyquist, 1932); multivariable feedback systems are then attacked via an appropriate generalisation of Bode's classical work on feedback theory (Bode, 1945). As some of the basic algebraic material presented is very recent, it will be developed in rather more detail than normal in a survey paper.

If we look back over the development of feedback control theory from our present vantage point, we can see that the chronological development falls into the four clearly defined phases shown in figure 1-1. Furthermore, the evolution of the theory shows a striking pattern which suggests that we are now approaching a point at which it should be possible to present a coherent and unified presentation of linear feedback theory. The object of this survey is to attempt such a task from the vector-frequency response viewpoint. The key concepts on which the whole of the treatment is based are the fundamental quantities of return-ratio and return-difference introduced by Bode. As a specific kind of development is being attempted, this will not lead to an exhaustive covering of all aspects of the theory. Nevertheless, it is hoped that all the key contributions to the subject have been treated and,

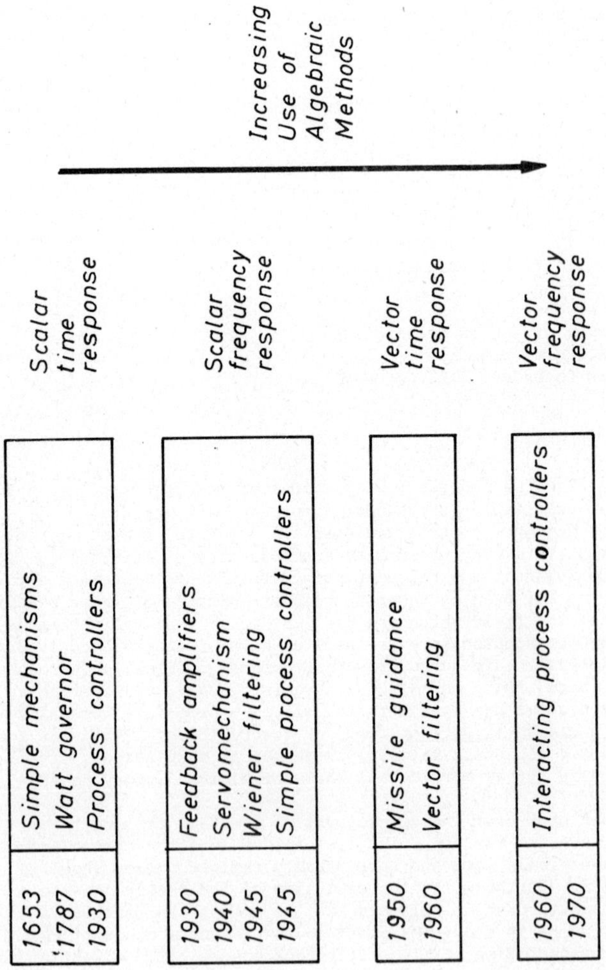

1653	Simple mechanisms	Scalar time response
1787	Watt governor	
1930	Process controllers	
1930	Feedback amplifiers	Scalar frequency response
1940	Servomechanism	
1945	Wiener filtering	
1945	Simple process controllers	
1950	Missile guidance	Vector time response
1960	Vector filtering	
1960	Interacting process controllers	Vector frequency response
1970		

Increasing Use of Algebraic Methods

Figure 1-1

by virtue of the use of the unifying concepts of return-ratio and return-difference, their inter-relationships indicated.

The survey is presented in what is hoped is a logical rather than a chronological order. An algebraic theory, based on Rosenbrock's work (Rosenbrock, 1967, 1970), is first sketched which defines the structural relationships in terms of which feedback systems may be manipulated into a variety of equivalent forms. The basic concepts of controllability and observability are introduced in this fundamental context. This is followed by stability theory, based on a vector generalisation of Nyquist's fundamental criterion, and the related concept of integrity. Bode's sensitivity results are then presented in a generalised vector form. The survey then continues with a sketch of the theory underlying all the main available techniques for the design of feedback systems. This starts with pole-shifting, followed by the techniques of optimal control and optimal filtering, and the use of observers for state reconstruction. The survey concludes with a review of multivariable feedback system design techniques from the frequency-response viewpoint.

2. ALGEBRAIC THEORY OF FEEDBACK SYSTEMS

A linear feedback system is here taken to be any dynamical system which may be represented by a signal-flow graph (or block diagram) of arbitrary complexity. The transmittances on the signal-flow graphs are functions of a complex variable s or z and relate Laplace transforms of signals for continuous-time systems or z-transforms of signals for discrete-time systems. The treatment given in this survey is mostly for continuous-time systems, since the algebraic relationships involved for both kinds of system are often essentially the same. In the following sections, specific results are quoted for both cases where appropriate; in this section only the continuous-time case is considered as the corresponding results for discrete-time systems simply involve, for the most part, replacing s by z.

The most important feature of an algebraic theory of feedback systems is that it enables us to discuss the <u>equivalence</u> theory of feedback systems. Thus, given any system with a stipulated overall transmittance between inputs and outputs, we can generate classes of equivalent systems having the <u>same</u> net transmittance between input and output. In particular, we can generate systems described by different, but equivalent, sets of equations associated with the system graph. This enables us to put any given system into a variety of standard forms, the most important of which is the <u>state</u> form associated with a set of first-order differential equations.

Since, given a signal-flow graph, we can construct an equivalent set of system equations, and vice versa, the theory can obviously be generated from a purely algebraic viewpoint without using specific graph representations. Such a treatment has been given by Rosenbrock (Rosenbrock, 1967, 1970). However, the use of graphs has two considerable advantages. The first, important here, is that lengthy algebraic manipulations are replaced by a simple appeal to graphical demonstrations. Secondly, the use of signal-flow graphs and block diagrams has become a most useful and widespread feature of feedback system analysis, and thus the treatment given should be more accessible to engineers than a purely algebraic version.

Suppose a general linear feedback system is represented by a signal-flow graph, as shown in the example of figure 2-1. Given such a graph, we can divide its signal vertices into three disjoint sets:

(i) input vertices,

(ii) output vertices,

(iii) all other vertices, which we may call internal, or dynamical, vertices.

The corresponding signal transform vectors associated with these vertex sets are:

(i) u(s), a vector of input signal transforms, of order l,

(ii) y(s), a vector of output signal transforms, of order m,

(iii) d(s), a vector of dynamical signal transforms, of order r.

Using these signal transform vectors, we may construct from any given signal-flow graph like 2-1 the equivalent vector signal-flow graph shown in figure 2-2. The four operator matrices U(s), V(s), T(s) and W(s) shown in figure 2-2 depend, of course, on the arbitrary order in which the various vertices are numbered within their selected sets. These four operators characterise the feedback system and may be given the following names:

(i) U(s) is the input operator matrix, of order r x l,

(ii) V(s) is the output operator matrix, of order m x r,

(iii) W(s) is the direct transference operator matrix, of order m x m,

(iv) T(s) is the return-ratio operator matrix, of order r x r for the selected set of dynamical vertices.

The term return-ratio arises from its use by Bode (Bode, 1945) in his fundamental investigations of feedback theory. Before we can outline the algebraic theory of feedback systems, one further operator matrix must be introduced. This may be introduced from a physical point of view in the following way, illustrated by figure 2-3. Suppose a disturbance vector n(s) of order r is injected (through a unit operator represented by I_r) on to the dynamical vertex set. Then we will have, on setting the inputs u(s) to zero that

$$d(s) = T(s)\, d(s) + n(s) \qquad (2\text{-}1)$$

so that

$$\left[I_r - T(s)\right] d(s) = n(s) \qquad (2\text{-}2)$$

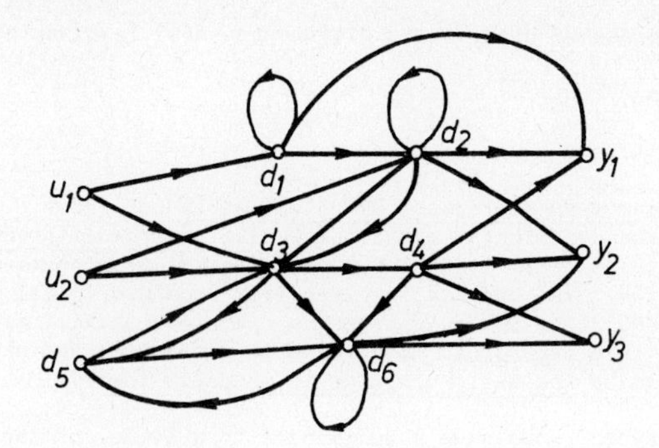

Figure 2-1

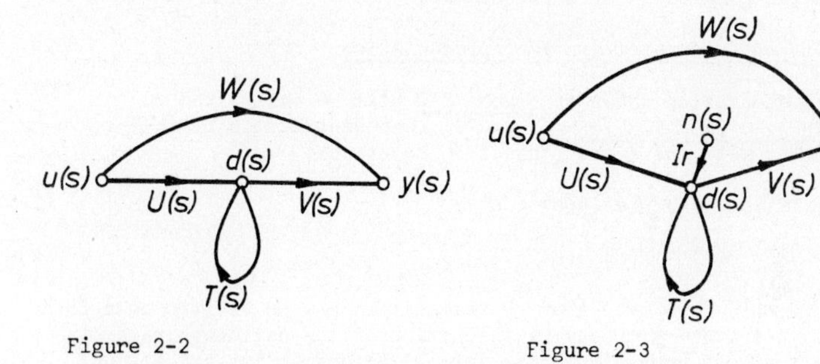

Figure 2-2

Figure 2-3

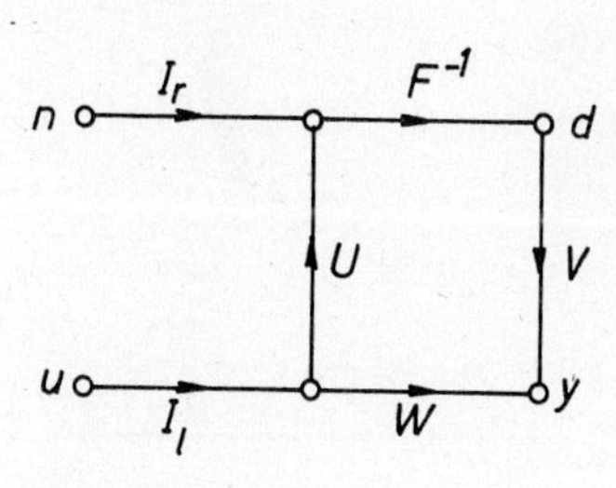

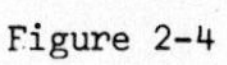
Figure 2-4

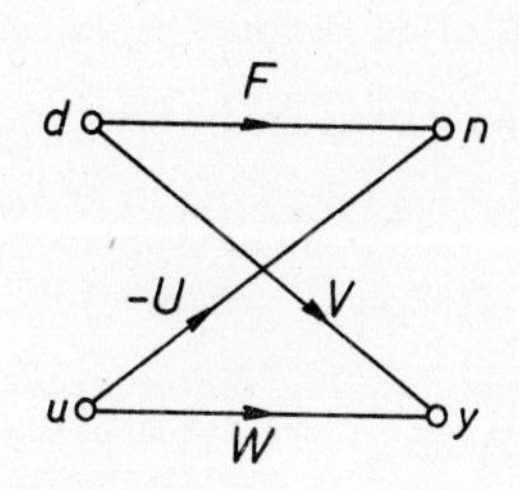

Figure 2-5

and thus the response of d(s) to the disturbance n(s) is given by

$$d(s) = F^{-1}(s)\, n(s) \qquad (2\text{-}3)$$

$$\text{where } F(s) = I_r - T(s) \qquad (2\text{-}4)$$

is defined as the return-difference operator matrix (of order r x r) associated with the dynamical variable set. The return-difference operator therefore determine the disturbance-rejection properties of the feedback system. The name again rises from Bode's original studies (Bode, 1945); the return-difference operator, as will be shown throughout this paper, has many valuable properties and can be reasonably said to be the basic operator in feedback theory.

In dealing with system models we usually have to provide for the effect of "initial conditions" in the variables of the associated differential (or difference) equations. This function can also be discharged by the extra input vector n(s) in figure 2-3 which can thus be regarded, as appropriate, as either representing the effects of initial conditions or disturbances.

2.1 Standard forms of system representation

The representation shown in figure 2-3 will be called the RETURN-RATIO OPERATOR FORM of system representation. It is associated with the set of transform equations:

$$d(s) = T(s)\, d(s) + U(s)\, u(s) + n(s) \qquad (2\text{-}5)$$

$$y(s) = V(s)\, d(s) + W(s)\, u(s) \qquad (2\text{-}6)$$

Although this form of system representation may be written down for any given system virtually by inspection of the defining graph, it is not the most convenient form for all algebraic investigations and we accordingly derive two related forms of representation from it by simple manipulations of the equation set (2-5). First we re-write equation (2-5) in the form

$$F(s)\, d(s) = U(s)\, u(s) + n(s) \qquad (2\text{-}7)$$

where F(s) is defined as in equation (2-4).

Thus, providing F(s) is not identically singular, we have

$$d(s) = F^{-1}(s)\, U(s)\, u(s) + F^{-1}(s)\, n(s) \qquad (2\text{-}8)$$

$$y(s) = V(s)\, d(s) + W(s)\, u(s) \qquad (2\text{-}9)$$

The corresponding vector signal-flow graph representation is shown in figure 2-4; this will be called the INVERSE RETURN-DIFFERENCE OPERATOR FORM of system representation.

Finally, re-arranging equation (2-7) and using equation (2-6) we get the set of equations

$$n(s) = F(s)\, d(s) - U(s)\, u(s) \qquad (2\text{-}10)$$

$$y(s) = V(s)\ d(s) + W(s)\ u(s) \qquad (2\text{-}11)$$

with the corresponding signal-flow graph representation shown in figure 2-5. This will be called the RETURN-DIFFERENCE OPERATOR FORM of system representation.

In the representation of figure 2-5, the disturbance vector n(s) appears as an output rather than an input. This re-arrangement of the normal causality is adopted for a limited and specific purpose in that it gives a form of representation in terms of which certain algebraic theory can be most effectively developed. For all other purposes the "normal causality" forms of figures 2-3 and 2-4 are used.

2.2 Equivalent Systems

For the systems considered above the output is given by

$$y(s) = \left[V(s)\ F^{-1}(s)\ U(s) + W(s)\right] u(s) + V(s)\ n(s) \qquad (2\text{-}12)$$

If n(s) is identically zero, input and output vectors are related by the transfer-function matrix

$$G(s) = V(s)\ F^{-1}(s)\ U(s) + W(s) \qquad (2\text{-}13)$$

Two feedback systems may be called equivalent if they have the same inputs and outputs and the same net transmittances between all input and output signal vertices, that is the same transfer function matrix G(s). A study of classes of equivalent feedback systems is of great importance since we frequently need to convert a given form of system representation to an equivalent representation of different structure. In this section we will consider, by simple arguments which use appropriate signal-flow graph representations directly, a set of system modifications which generate equivalent systems. These system modifications are then re-interpreted in algebraic terms. In all the arguments which follow, it is explicitly assumed that det F(s) does not vanish identically in s, and thus that F(s) is invertible. We will also, where convenient, drop the indication of explicit dependence on s.

First consider the simultaneous changes

$$U \to U + FY \text{ and } W \to W - VY$$

where Y(s) is some arbitrarily chosen matrix of appropriate dimensions. An inspection of the inverse return-difference form of representation shown in figure 2-6 shows that the consequent net change in transmission between input u(s) and output y(s) is

$$VY - VY = 0$$

so that every modification of this sort generates an equivalent feedback system. To find an algebraic description of such an equivalence operation we write the appropriate equations in the return-difference operator form which gives

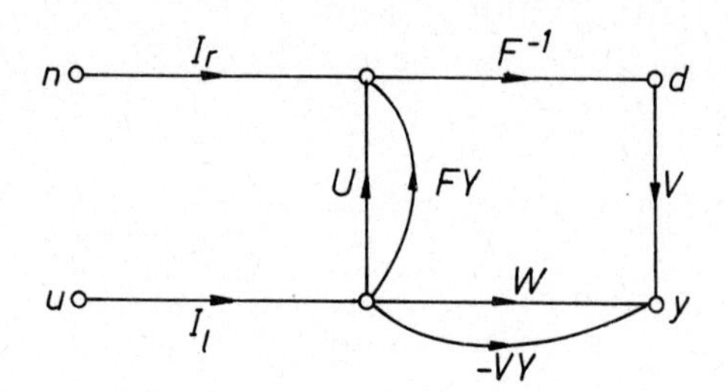

Figure 2-6

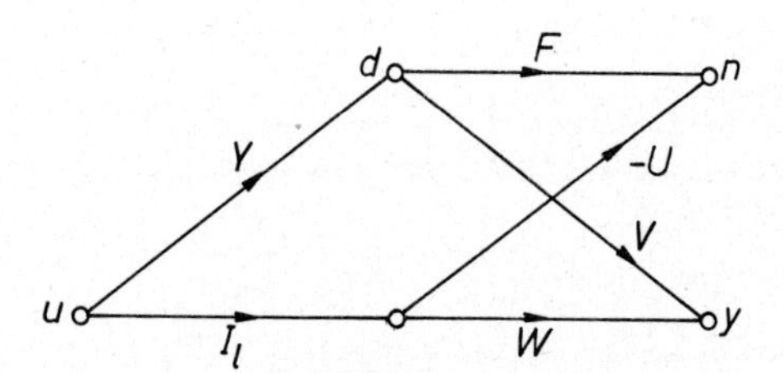

Figure 2-7

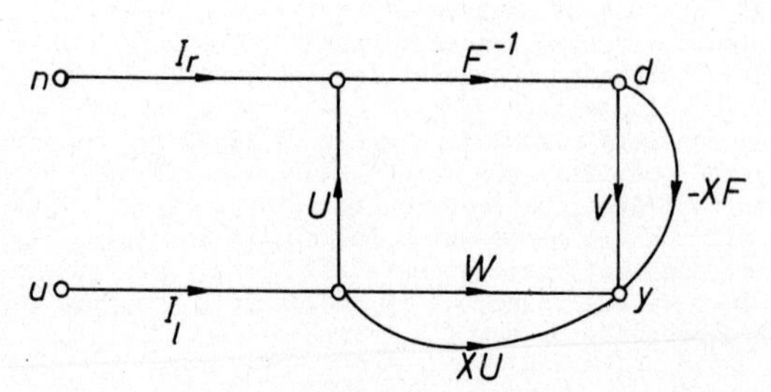

Figure 2-8

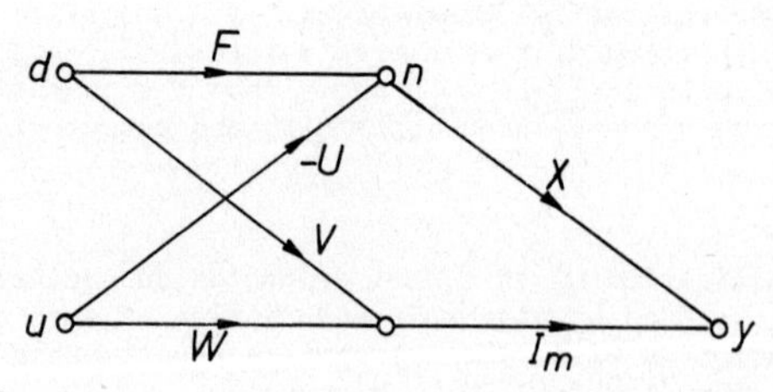

Figure 2-9

$$n = Fd - (U + FY)u$$
$$y = Vd + (W - VY)u \qquad (2\text{-}14)$$

which may be written in matrix form as

$$\begin{pmatrix} n \\ y \end{pmatrix} = \begin{pmatrix} F & -U \\ V & W \end{pmatrix} \begin{pmatrix} I_r & Y \\ 0 & I_l \end{pmatrix} \begin{pmatrix} d \\ u \end{pmatrix} \qquad (2\text{-}15)$$

This transformation may be represented by the signal-flow graph shown in figure 2-7.

2.2.1 System Matrix

The implications of equation (2-15) are very important: it shows that both a system and an equivalence operation on a set of systems may be represented in matrix terms. This shows the possibility of representing a system as an algebraic entity, which immediately leads to an appropriate algebraic system theory. The system may be represented, for the purposes of discussing equivalence in algebraic terms, by the matrix

$$P(s) = \begin{pmatrix} F(s) & -U(s) \\ V(s) & W(s) \end{pmatrix} \qquad (2\text{-}16)$$

which, following Rosenbrock (Rosenbrock 1967, 1970), we will call the SYSTEM MATRIX. Representing systems in this way (except for a minor notational and sign change), Rosenbrock has developed a very complete and powerful algebraic system theory (Rosenbrock, 1970), which we now proceed to apply to feedback systems. Note at this point that, given any system matrix, we can immediately construct a system representation, in signal-flow graph terms or standard equation terms, and vice versa. Note also that equation (2-15) shows that equivalent system matrices are generated by an operation easily represented in algebraic terms as post-multiplication by an appropriate matrix. These observations show how an algebraic theory of equivalent feedback systems may be developed.

Proceeding along these lines we now consider the simultaneous changes

$V \rightarrow V - XF$ and $W \rightarrow W + XU$

where X is an arbitrary matrix of suitable dimensions. An inspection of the representation shown in figure 2-8 shows that the net input-output transmission change is

$$- XU + XU = 0$$

so that such modifications again generate equivalent systems. The appropriately modified equations in this case may be written in matrix form as

$$\begin{pmatrix} n \\ y \end{pmatrix} = \begin{pmatrix} I_r & 0 \\ X & I_m \end{pmatrix} \begin{pmatrix} F & -U \\ V & W \end{pmatrix} \begin{pmatrix} d \\ u \end{pmatrix} \tag{2-17}$$

corresponding to the signal-flow graph of figure 2-9.

Finally, we may modify the system of figure 2-4 to give the system shown in figure 2-10 in which M and N are arbitrary invertible matrices of appropriate dimensions. An inspection of the signal-flow graphs shows that the two systems are equivalent, and simple manipulation of the corresponding equation sets shows that the equivalence may be expressed in algebraic form by the matrix equation

$$\begin{pmatrix} \bar{n} \\ y \end{pmatrix} = \begin{pmatrix} M & 0 \\ I_r & I_m \end{pmatrix} \begin{pmatrix} F & -U \\ V & W \end{pmatrix} \begin{pmatrix} N & I_r \\ 0 & I_l \end{pmatrix} \begin{pmatrix} \bar{d} \\ u \end{pmatrix} \tag{2-18}$$

corresponding to the graph of figure 2-11.

The discussion so far has shown that the system matrix for an equivalent system is generated by pre- or post-multiplication by an appropriate matrix. All the equivalence operations so far considered may be combined into the composite equivalence operation represented in signal-flow graph form in figure 2-12, with corresponding matrix representation

$$\begin{pmatrix} \bar{n} \\ y \end{pmatrix} = \begin{pmatrix} M & 0 \\ X & I_m \end{pmatrix} \begin{pmatrix} F & -U \\ V & W \end{pmatrix} \begin{pmatrix} N & Y \\ 0 & I_l \end{pmatrix} \begin{pmatrix} \bar{d} \\ y \end{pmatrix} \tag{2-19}$$

Two further types of system modification must be considered in order to give a complete equivalence theory. We first note that if a system graph (in inverse return-difference form) is drawn in partitioned form for sets of dynamical vertices split up into groups as shown in figure 2-13, then obviously the re-arranged systems obtained by interchanging the positions of the vertices are, trivially, equivalent. The algebraic interpretation of this is noted in the summary of algebraic equivalence operations given below.

A much more interesting equivalence operation is the addition of deletion of uncontrollable or unobservable systems as shown in figure 2-14 and figure 2-15. (The terms uncontrollable and unobservable are again considered in a different and more algebraically technical way below when state space representations are considered). The presence of an uncontrollable sub-system, as shown in figure 2-14, gives rise to a system matrix of the form

$$\begin{pmatrix} F_{11} & 0 & 0 \\ F_{21} & F_{22} & -U_2 \\ V_1 & V_2 & W \end{pmatrix}$$

The presence of an unobservable sub-system, as shown in figure 2-15,

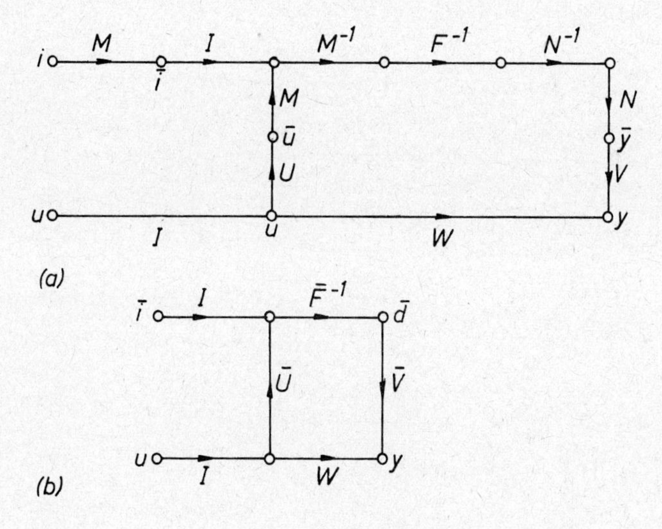

Figure 2-10

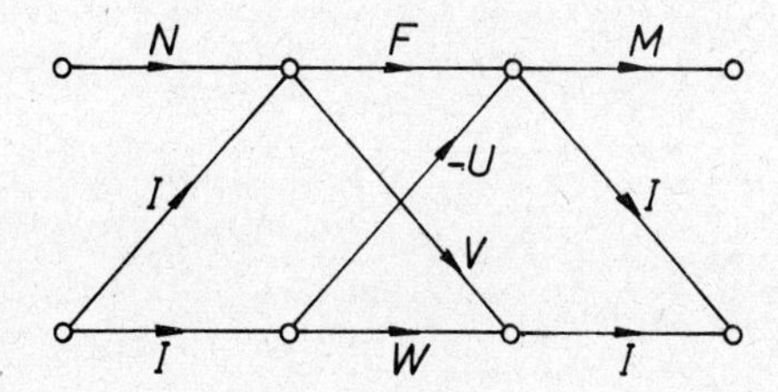

Figure 2-11

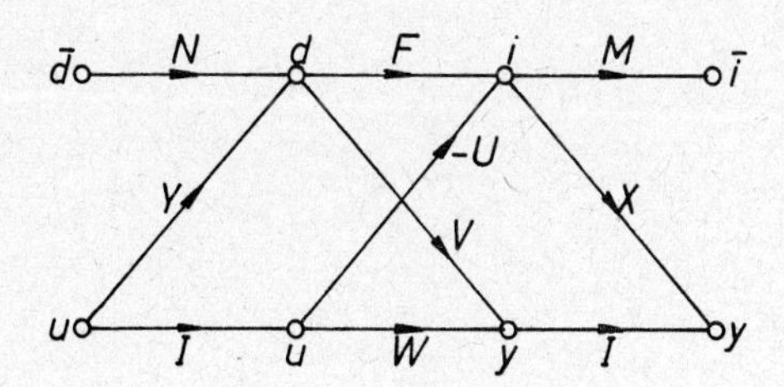

Figure 2-12

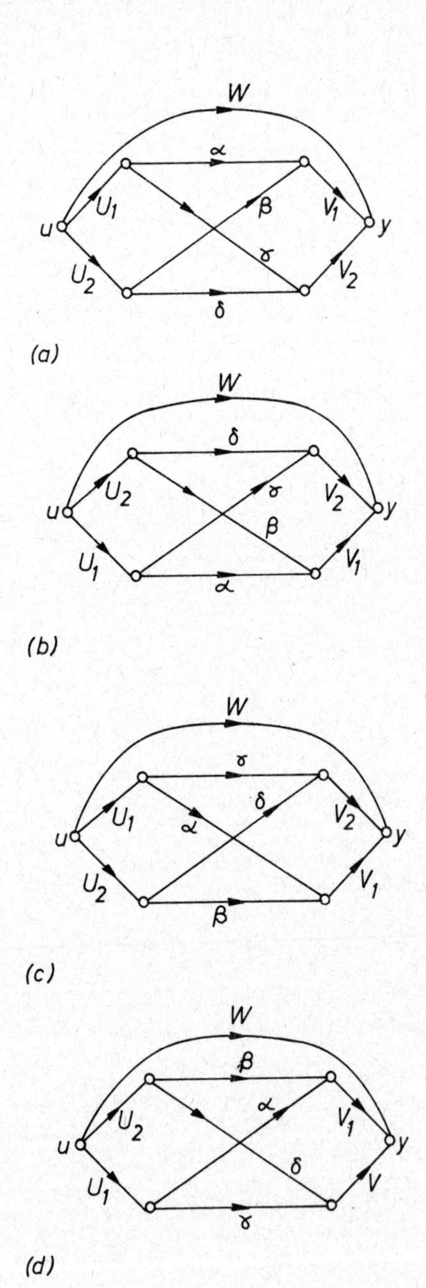
W
α
U1
β
V1
u
y
γ
U2
V2
δ
(a)
W
δ
U2
γ
V2
u
y
β
U1
V1
α
(b)
W
γ
δ
U1
α
V2
u
y
U2
V1
β
(c)
W
β
α
U2
V1
u
y
δ
U1
V
γ
(d)

Figure 2-13

gives rise to a system matrix of the form

$$\begin{pmatrix} F_{11} & F_{12} & -U_1 \\ 0 & F_{22} & -U_2 \\ 0 & V_2 & W \end{pmatrix}$$

In general, any system may have its dynamical vertices grouped in such a way that they fall into the following four disjoint sets, as illustrated in figure 2-16:

(i) not controllable and not observable, corresponding to vertex d_1

(ii) controllable and not observable, corresponding to vertex d_2

(iii) observable and not controllable, corresponding to vertex d_3

(iv) controllable and observable, corresponding to vertex d_4.

The associated system matrix has the form

$$\begin{pmatrix} F_{11} & 0 & F_{13} & 0 & 0 \\ F_{21} & F_{22} & F_{23} & F_{24} & U_2 \\ 0 & 0 & F_{33} & 0 & 0 \\ 0 & 0 & F_{43} & F_{44} & U_1 \\ 0 & 0 & V_3 & V_1 & W \end{pmatrix}$$

2.2.2 Summary of Algebraic Equivalence Operations

Equivalent systems are generated from any given system matrix by a set of standard operations on the system matrix which may be summarised as below

(i) Any of the first r rows (or columns) may be multiplied by a rational function, not identically zero.

(ii) Any multiple of a row (or column) by a rational function may be added to any other row (or column).

(iii) Any two rows (or columns) may be interchanged.

(iv) The matrix may be augmented or reduced in the manner appropriate to the addition or deletion of uncontrollable or unobservable sub-systems.

2.2.3 Standard Forms of System Matrix

Certain specific kinds of system representation are associated with specific forms of system matrix, and it is convenient to note some of them at this point.

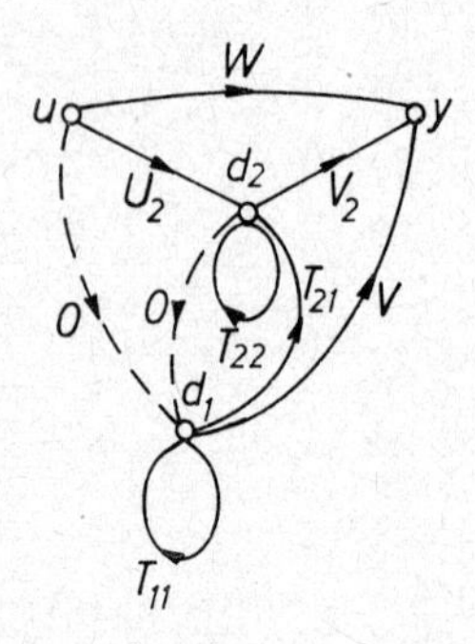

Figure 2-14

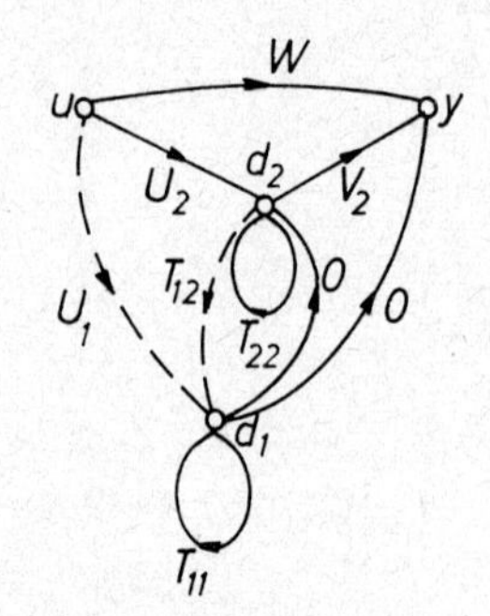

Figure 2-15

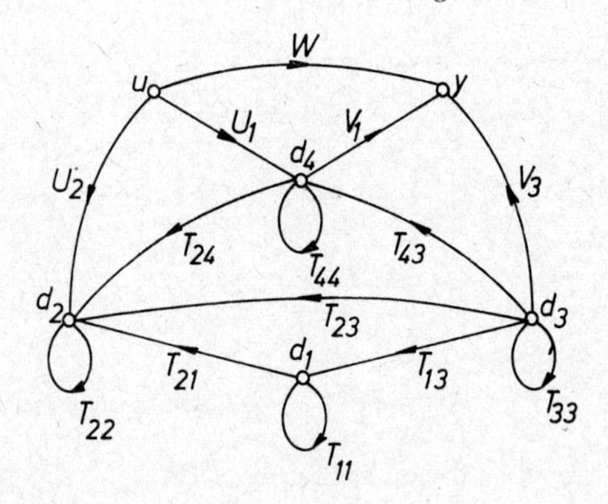

Figure 2-16

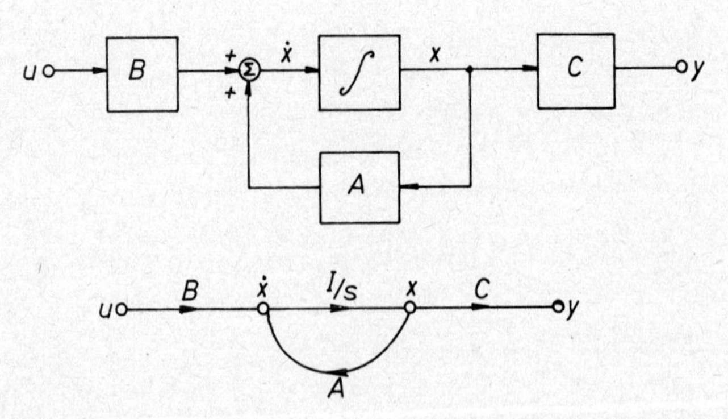

Figure 2-17

Rational Form. - If all the entries in the system matrix are rational functions in s, it is said to be of rational form.

Polynomial Form. - Using equivalence operations of the kind discussed above, Rosenbrock (Rosenbrock, 1970) has shown that any system matrix can be put into a form in which all its entries are polynomials in s; such a system matrix is said to be in polynomial form.

This form of system matrix has great practical importance since it associates the feedback system with a set of equations

$$\begin{aligned} F(s)\, d(s) &= U(s)\, u(s) + n(s) \\ y(s) &= V(s)\, d(s) + W(s)\, u(s) \end{aligned} \qquad (2\text{-}20)$$

in which F(s), U(s), V(s) and W(s) are all polynomial matrices. This transformed set of equations (2-20) is therefore related to a set of linear constant coefficient differential equations for the system. It immediately follows from this that the characteristic polynomial of the system is det F(s) (where F(s) is the polynomial matrix in equation (2-20)) and that the order of the dynamical system is the order of the polynomial det F(s).

State Form.- It can be further shown that, again using equivalence operations of the type discussed above, any system matrix can be put into the form

$$\begin{pmatrix} (sI - A) & -B \\ C & D(s) \end{pmatrix}$$

where A, B, C are matrices of real numbers and D(s) is a polynomial matrix; such a system matrix is said to be in state space form. If D(s) is zero the system is said to be proper; this has the physical significance of implying a frequency response which becomes arbitrarily small for arbitrarily large values of frequency. The state form of system matrix associates the feedback system with the set of transformed equations

$$\begin{aligned} sx(s) &= Ax(s) + Bu(s) \\ y(s) &= Cx(s) + D(s)\, u(s) \end{aligned} \qquad (2\text{-}21)$$

where, according to a well-established convention for state variables, x(s) has been used to denote the dynamical variables. For a proper system we immediately obtain from equation (2-21) the associated system state-space differential equation set

$$\begin{aligned} \frac{dx}{dt} &= Ax + Bu \\ y &= Cx \end{aligned} \qquad (2\text{-}22)$$

corresponding to the graph representations of figure 2-17.

2.3 Controllability and Observability

Suppose we have a system for which an associated polynomial form of system matrix

$$P(s) = \begin{pmatrix} F(s) & -U(s) \\ V(s) & W(s) \end{pmatrix}$$

is such that there exists a polynomial matrix Q(s) such that

$$F(s) = Q(s)\, F_1(s) \text{ and } U(s) = Q(s)\, U_1(s) \tag{2-23}$$

F(s) and U(s) are said to have the common left divisor Q(s). We can then relate the original system matrix to a new equivalent system matrix by

$$\begin{pmatrix} F & -U \\ V & W \end{pmatrix} = \begin{pmatrix} Q & 0 \\ 0 & I_m \end{pmatrix} \begin{pmatrix} F_1 & -U_1 \\ V & W \end{pmatrix} \tag{2-24}$$

Further, if F(s) and V(s) also have a common right divisor polynomial matrix R(s) so that

$$F = (QF_1)R \quad \text{and} \quad V = V_1R \tag{2-25}$$

then we will have

$$\begin{pmatrix} F & -U \\ V & W \end{pmatrix} = \begin{pmatrix} Q & 0 \\ 0 & I_m \end{pmatrix} \begin{pmatrix} F_1 & -U_1 \\ V_1 & W \end{pmatrix} \begin{pmatrix} R & 0 \\ 0 & I_1 \end{pmatrix} \tag{2-26}$$

Equivalence may be verified by forming the transfer function matrix

$$G(s) = VF^{-1}U + W = V_1R\,(QF_1R)^{-1}\,QU_1 + W = V_1F_1^{-1}U_1 + W \tag{2-27}$$

which shows how the common divisors cancel to give an unaltered transfer function matrix (Rosenbrock, 1970).

Reduction of System Order. - An inspection of equation (2-26) shows that det $F_1(s)$ will have a lower order than det F(s) so that the removal of common factors in this way generates an equivalent system of reduced order. A systematic application of this technique gives the least order representation for any given system.

The basic algebraic theory of linear dynamical system structure was first developed by Kalman in a classic set of papers. (Kalman, 1960, 1962, 1963, 1965). These attacked the problem from the state-space point of view and developed the key concepts of controllability, observability, canonical representation and minimal order realization. Kalman also gave (Kalman, 1965) a most important study of the relationship between minimal order representations and the degree of a rational matrix. This used the classical study of McMillan on the role of canonical forms of rational matrices in realizability theory (McMillan, 1952).

2.3.1 Controllability

A system in state-space form with associated differential equation set

$$\frac{dx}{dt} = Ax + Bu \tag{2-28}$$

is said to be controllable if, given any initial state $x(o) = C$, there exists a time $t_1 > o$ and a control u defined on $[o,t_1]$ such that $x(t_1) = o$.

For a discrete-time system, with difference equation set

$$x_{k+1} = Ax_k + Bu_k \tag{2-29}$$

the system is said to be controllable if there exists an integer $p > o$ and a sequence $u_o, u_1, ---, u_{p-1}$ such that $x_p = o$.
Rosenbrock has shown that the system (2-28) is controllable if and only if $sI - A$ and B are relatively left prime (that is have no common left factors other than a unit matrix), and that the discrete system (2-29) is controllable if and only if $zI - A$ and B are relatively left prime.

A classic study of controllability from the state-space point of view was given by Kalman, Ho and Narendra (Kalman, Ho, Narendra, 1961). This established the basic state-space pointwise-state controllability theorem that the system of equations (2-28) will be controllable if and only if the matrix

$$\left[B \; AB \; A^2B \; ---- \; A^{n-1}B\right]$$

has rank n. This result was also established in a simple way in an important paper by Gilbert (Gilbert, 1963) which gave a valuable discussion of the links between state-space and transfer-function-matrix representations of system behaviour.

2.3.2 Observability

A system in state-space form with identically zero input represented by the state-space equations

$$\begin{aligned} \dot{x} &= Ax \\ y &= Cx \end{aligned} \tag{2-30}$$

is said to be observable if there exists a $t_1 > o$ such that given y on the interval $[o, t_1]$ it is possible to deduce $x(o)$. For the discrete system

$$\begin{aligned} x_{k+1} &= Ax_k \\ y_k &= Cx_k \end{aligned} \tag{2-31}$$

the system is called observable if there exists a $q > o$ such that given $y_o, y_1 ----, y_{q-1}$ it is possible to deduce x_o.

The system (2-30) is observable if and only if sI - A and C are relatively right prime. The discrete case follows by sustituing z for s.

Observability is fully discussed from the state-space viewpoint in the papers by Kalman, Ho and Narendra and Gilbert cited above in the section on controllability. In these papers the basic state-space observability criterion was established that the system of equations (2-30) is observable if and only if the matrix

$$\left[C^t \; A^tC^t \; - - - - \; (A^t)^{n-1} \; C^t\right]$$

has rank n where t denotes transposition.

2.4 Extension of Definitions of Return-Ratio and Return-Difference

In what follows it is useful to have an extended definition of return-ratio and return-difference. It will be appreciated later that this could be achieved using the equivalence theory just outlined, but it is nevertheless useful to introduce it at this point in a simple way.

Given any feedback system, such as that shown in figure 2-1, we can select any specific sub-set of the dynamical vertex set and eliminate all other dynamical vertices using normal signal-flow graph manipulation techniques. This will, of course, introduce extra direct transmission paths between input and output. We may then form the vector representation of figure 2-2 for this reduced system and refer to the corresponding matrix T(s) as the return-ratio matrix *for the specific selected vertex set*. The corresponding return-difference matrix for the specific selected vertex set is then defined in the usual way.

3. STABILITY

The feedback configuration which is most often studied in connection with practical applications of feedback theory in automatic control studies is shown in figure 3-1, in which

r(s) = m x 1 matrix of reference input transforms
e(s) = m x 1 matrix of error transforms
y(s) = m x 1 matrix of plant output transforms
u(s) = r x 1 matrix of plant input transforms
K(s) = r x m matrix of controller transfer functions
G(s) = m x r matrix of plant transfer functions
H(s) = m x m matrix of feedback-transducer transfer functions

By simple signal - flow graph manipulations, it can be replaced by any of the three equivalent systems shown in figures 3-2, 3-3 and 3-4, for which the following quantities are defined:

$$-G(s)K(s)H(s) = T_y(s) = \text{return-ratio matrix for output vertex} \quad (3\text{-}1)$$

$$-K(s)H(s)G(s) = T_u(s) = \text{return-ratio matrix for plant input vertex} \quad (3\text{-}2)$$

$$-H(s)G(s)K(s) = T_e(s) = \text{return-ratio matrix for error vertex} \quad (3\text{-}3)$$

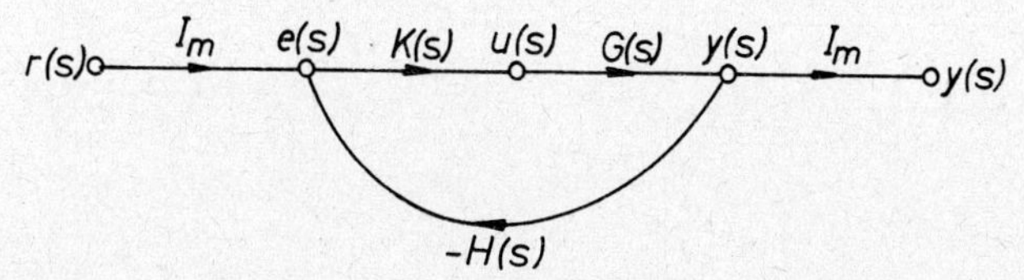

Figure 3-1

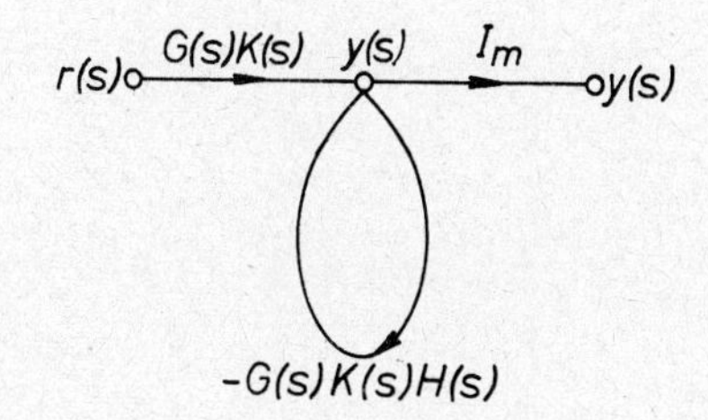

Figure 3-2

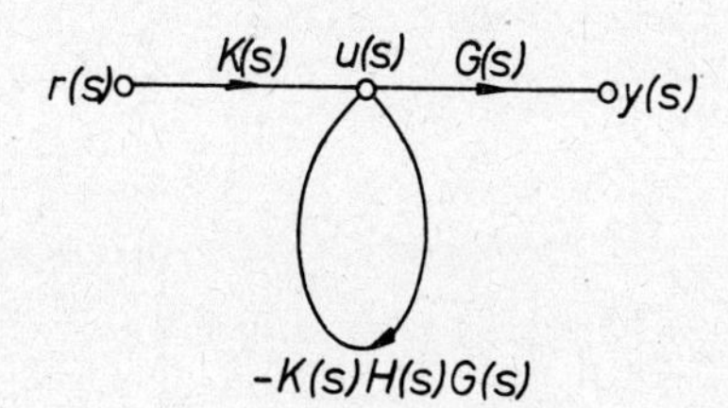

Figure 3-3

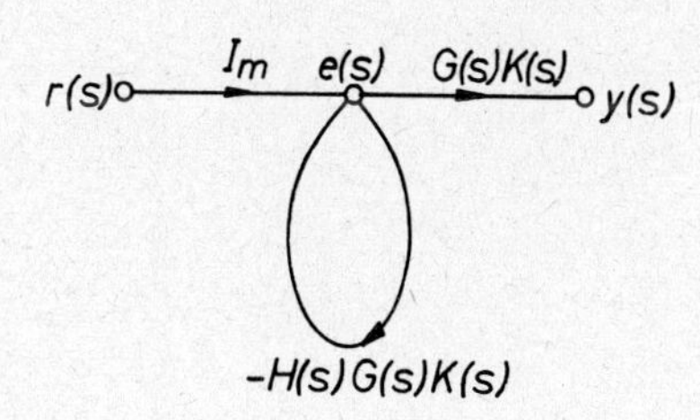

Figure 3-4

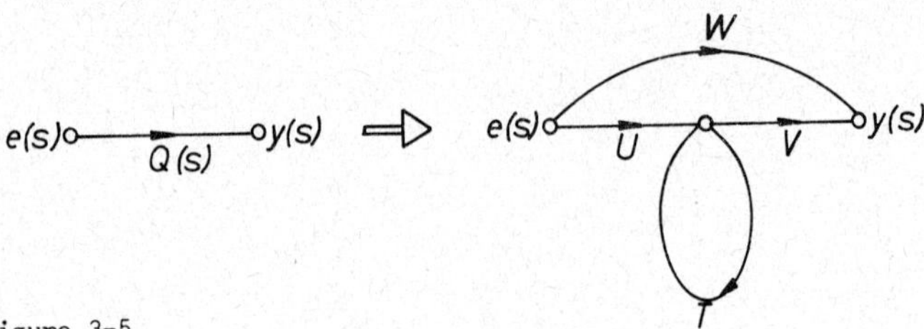

Figure 3-5

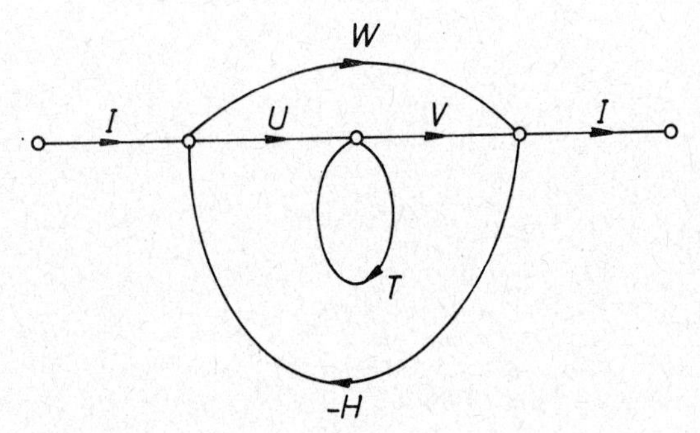

Figure 3-6

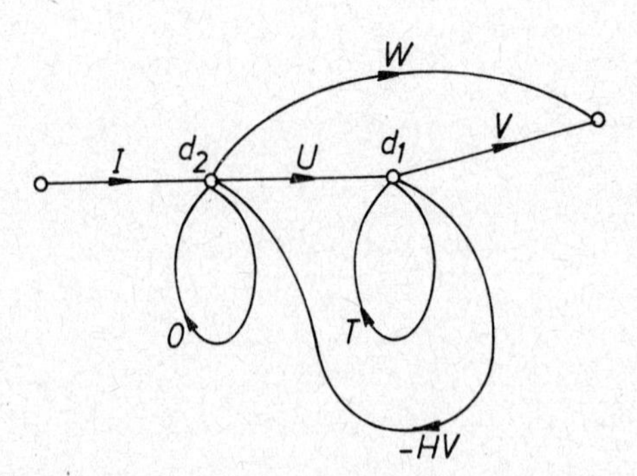

Figure 3-7

The corresponding return-difference matrices for each of these three vertices are defined as

$$F_y(s) = \left[I_m + G(s)K(s)H(s)\right] \quad (3\text{-}4)$$

$$F_u(s) = \left[I_r + K(s)H(s)G(s)\right] \quad (3\text{-}5)$$

$$F_e(s) = \left[I_m + H(s)G(s)K(s)\right] \quad (3\text{-}6)$$

Simple algebraic manipulations give that, in terms of these return-ratio matrices, the closed-loop transfer function matrix relating r(s) and y(s) in figure 3-1 is given by the following set of equivalent expressions

$$R(s) = F_y^{-1}(s)G(s)K(s) = G(s)F_u^{-1}(s)K(s) = G(s)K(s)F_e^{-1}(s) \quad (3\text{-}7)$$

It is convenient in the rest of this survey to put

$$G(s)K(s) = Q(s) \quad (3\text{-}8)$$

and we will frequently refer to Q(s) as the system open-loop transfer function matrix.

Now suppose we start with an open-loop system Q(s) and, as shown in figure 3-5, replace it by an equivalent feedback system in which all transferences are polynomial matrices in s; this may be done by using the equivalence operations outlined in Section 2. If, as usual, we introduce the corresponding return-difference matrix F(s) = I - T(s) then we can take det F(s) as the characteristic polynomial of the open-loop system, that is of the system represented by Q(s). Thus

$$\text{open-loop characteristic polynomial} = \text{O.L.C.P.} = \det F(s) \quad (3\text{-}9)$$

If we now complete the system of figure 3-1 by adding the feedback represented by H(s), we get the system shown in figure 3-6, which we can then re-draw in the equivalent representation shown in figure 3-7. Using the definition of system matrix given in Section 2 above we can now write down the system matrix for the feedback system shown in figure 3-7 as

$$P(s) = \left(\begin{array}{cc|c} F & U & 0 \\ -HV & I & I \\ \hline V & 0 & W \end{array}\right) \quad (3\text{-}10)$$

and, assuming for simplicity that H(s) is a constant matrix (or a polynomial matrix) we have that

$$\text{closed-loop characteristic polynomial} = \text{C.L.C.P.}$$

$$= \det \begin{pmatrix} F & U \\ -HV & I \end{pmatrix} \quad (3\text{-}11)$$

Now, using Schur's formula for partitioned determinants (Gantmacher, 1959) we have that

$$\det\begin{pmatrix} F & U \\ -HV & I \end{pmatrix} = \det F \det (I + HVF^{-1}U)$$

$$= \det F \det (I + HQ) \qquad (3\text{-}12)$$

and thus, from equations (3-9), (3-11) and (3-12) we get the <u>fundamental equation relating open- and closed-loop system characteristic polynomials</u>

$$\det F_y(s) = \frac{\text{C.L.C.P.}}{\text{O.L.C.P.}} \qquad (3\text{-}13)$$

A further simple application of Schur's formulae shows that

$$\det F_y(s) = \det F_u(s) = \det F_e(s) \qquad (3\text{-}14)$$

so that it is, in this sense, immaterial which of the three return-difference matrices we use in equation (3-13).

3.1 Fundamental Multivariable Feedback Stability Theorem

We may now use equation (3-13) to derive a fundamental stability theorem for multivariable systems in terms of $\det F_y(s)$. Let D be a contour in the complex plane described clockwise and consisting of the imaginary axis from $-j\alpha$ to $+j\alpha$ where α is the radius of a large semi-circle in the right-half complex plane which is large enough to enclose every zero of $\det Q(s)$ and $\det R(s)$ which lies within D, and is centred on the origin. Then it has been shown by Rosenbrock and McMorran (Rosenbrock, 1970), (Rosenbrock, 1971), (McMorran, 1970), that if $\det R(s)$ maps D into Γ_c and encircles the origin n_c times clockwise, and $\det Q(s)$ maps D into Γ_o encircling the origin n_o times clockwise, then the multivariable feedback system is closed-loop stable if and only if

$$n_c - n_o = p_o \qquad (3\text{-}15)$$

where p_o is the number of right-half-plane zeros of the open-loop characteristic polynomial. Since

$$\det F_y(s) = \frac{\det Q(s)}{\det R(s)} \qquad (3\text{-}16)$$

it follows immediately that if $\det F_y(s)$ maps D into Γ_f, encircling the origin of the complex plane n_f times in a clockwise direction then

$$n_f = n_o - n_c \qquad (3\text{-}17)$$

It is therefore evident from equations (3-15) and (3-17) that a necessary and sufficient condition for closed-loop stability is that

$$n_f = - p_o \qquad (3\text{-}18)$$

This is of great practical importance since standard Nyquist stability techniques (Nyquist, 1932) can now be applied to a complex plane mapping of det $F_y(s)$ which will ensure closed-loop stability if and only if the origin encirclement requirement is satisfied. The open-loop stable case follows simply on setting p_o equal to zero.

Stability criteria in terms of system return-difference determinants do not provide enough information for a flexible design technique and the fundamental stability criteria are therefore extended by introducing the concept of characteristic transfer functions (MacFarlane, 1970) for square matrices whose elements are transfer functions, that is for square transfer function matrices. It is important to note at this point that, by definition, all return-ratio and return-difference matrices are square and will thus, in general, have eigenvalues in the field over which they are defined.

3.2 Characteristic Transfer Functions

All the matrices considered here, such as G(s), K(s), H(s) $T_e(s)$, $T_y(s)$, $T_u(s)$, $F_e(s)$, $F_y(s)$, $F_u(s)$, etc., are matrices whose elements are contained in the field of rational functions. Such matrices may be discussed within the general framework of the theory of linear operators on vector spaces over fields (MacFarlane, 1970). Any m-dimensional vector space over the field of rational functions in s introduced for this purpose can be thought of as spaces in which reside m-dimensional vectors of signal transforms, each transform (or vector component) being a rational fraction in s. An operator on such a space, represented by some matrix Q(s) say, will convert one vector of signal transforms, such as a system-input transform set, into another vector of signal transforms, such as a system-output transform set. If $\omega(s)$ is an eigenvector of Q(s) with corresponding eigenvalue q(s) then

$$Q(s)\omega(s) = q(s)\omega(s) \tag{3-19}$$

which shows that the action of Q(s) on $\omega(s)$ is such that every transform component of $\omega(s)$ is multiplied by the same scalar transfer function q(s). The transfer function q(s) is therefore called a characteristic transfer function of Q(s) with an associated characteristic frequency response $q(j\omega)$.

In general, the eigenvalues of a matrix of rational functions, regarded as an operator on a finite-dimensional vector space over the field of rational functions, may not lie in the field of rational functions. They will frequently be irrational functions. This causes no difficulty in practice since any calculations necessary for design purposes will normally be carried out for specific values of s (usually distributed along the imaginary axis in the complex plane) and all computations are then carried out over the field of complex numbers. From a theoretical point of view this difficulty is overcome by following the standard algebraic practice of defining a suitable extension field within which the field of rational functions is embedded. The transfer-function matrices are then regarded as representing operators over the extension field.

3.3 Stability in Terms of Characteristic Frequency Response Loci

The results presented here hold for any of the three sets of return-ratio and return-difference matrices $T_e(s)$, $F_e(s)$, $T_u(s)$ $F_u(s)$ and $T_y(s)$, $F_y(s)$ defined above. They will accordingly be presented for some general return-difference matrix F(s), of dimension m x m, with an associated return-ratio matrix T(s).

Let the return-difference matrix F(s) have the set of eigenvalues, or characteristic transfer functions

$\rho_j(s) \quad j = 1,2,\ldots,m$

Let $\rho_j(s)$ map D into Γ_{fj} $(j = 1,2,\ldots,m)$ where each locus Γ_{fj} encircles the origin in the complex plane n_{fj} times in a clockwise direction.

Then (MacFarlane and Belletrutti, 1971)

$$\sum_{j=1}^{m} n_{fj} = \text{sum of origin encirclements for all } \rho_j(s) \qquad (3\text{-}20)$$

Now, since

$$\det F(s) = \prod_{j=1}^{m} \rho_j(s) \qquad (3\text{-}21)$$

it follows that

$$\sum_{j=1}^{m} n_{fj} = n_f \qquad (3\text{-}22)$$

and thus closed-loop stability is assured if and only if

$$\sum_{j=1}^{m} n_{fj} = -p_o \qquad (3\text{-}23)$$

A similar criterion can be developed for return-ratio matrices. Since we have that

$$F(s) = I_m + T(s) \qquad (3\text{-}24)$$

the eigenvalue-shift theorem gives that

$$\rho_j(s) = 1 + \gamma_j(s) \quad j = 1,2,\ldots,m \qquad (3\text{-}25)$$

where $\{\gamma_j(s)\ ;\ j = 1,2,\ldots,m\}$ are the characteristic transfer functions of T(s). The effect of equation (3-25) in terms of complex-plane mappings of $\gamma_j(s)$ is simply to shift the critical point from the origin to the usual (-1,0) location. Thus, letting $\gamma_j(s)$ map D into Γ_{tj} $(j = 1,2,\ldots,m)$ encircling the (-1,0) point n_{tj} times clockwise, it follows that from equation (3-25) that

$$n_{fj} = n_{tj} \tag{3-26}$$

and it is therefore necessary and sufficient for closed-loop stability that

$$\sum_{j=1}^{m} n_{tj} = -p_o \tag{3-27}$$

To sum up, the closed-loop system is stable if and only if the net sum of critical point encirclements of all the characteristic loci $\gamma_j(j\omega)$ for $j = 1,2,\ldots,m$ is counter-clockwise and equal to the number of rhp zeros of the open-loop system characteristic polynomial. For the open-loop stable case, no characteristic frequency response $\gamma_j(j\omega)$ must enclose the (-1,0) point in the complex plane.

3.4 Multivariable Root Locus Techniques

Equation (3-13) shows how the roots of the closed-loop system characteristic equation depend on det F(s) (where F(s), as above, can be any of $F_e(s)$, $F_u(s)$ or $F_y(s)$ as convenient). Consider the particular case of $F_y(s)$

$$\det F_y(s) = \det\left[I_m + G(s)K(s)H(s)\right] \tag{3-28}$$

and suppose we put

$$G(s)K(s)H(s) = k\, Q(s) \tag{3-29}$$

where k is a variable scalar gain-multiplier and Q(s) is a fixed transfer-function matrix. We may then expand the determinant on the right-hand side of equation (3-28) to give (Retallack, 1970)

$$\det\left[I_m + k\, Q(s)\right]$$

$$= 1 + k \operatorname{trace} Q + k^2 \sum (\text{principal minors of order 2 of } Q)$$

$$+ - - - - + k^m \det Q \tag{3-30}$$

If this is now equated to zero we get an equation which is a natural generalization of the basic equation used in the scalar root locus method. This has been developed into a multivariable root-locus method by Retallack (Retallack, 1970).

4. INTEGRITY

In designing a feedback controller for a multivariable system it is important to check the stability of the set of systems which result when one or more transducers or actuators fail. A system which remains stable under all likely failure conditions is said to be a high integrity system. The stability theory of the previous section has been extended to include such contingencies by Belletrutti and MacFarlane (Belletrutti and MacFarlane, 1971), who have shown that a multivariable feedback system of the type shown in figure 3-1 will have high integrity if, and only if, all the characteristic loci of all the principal submatrices of the return-ratio matrices $T_e(s)$,

$T_u(s)$ and $T_y(s)$ satisfy Nyquist stability criteria as defined by equation (3-27).

5. SENSITIVITY

Bode's classical studies (Bode, 1945) involved the characterisation of sensitivities, as well as stability, in terms of return-differences. The treatment of sensitivity properties of multivariable systems which is sketched here is based on work done by McMorran (McMorran, 1970).

Consider the standard feedback configuration shown in figure 3-1. The open-loop transfer-function matrix is given by

$$Q(s) = G(s)K(s) \tag{5-1}$$

and the closed-loop transfer-function matrix is

$$R(s) = \left[I_m + Q(s)H(s)\right]^{-1} Q(s) \tag{5-2}$$

Thus, since Q(s) is generally a square matrix, we have

$$R^{-1}(s) = Q^{-1}(s)\left[I_m + Q(s)H(s)\right] = Q^{-1}(s) + H(s) \tag{5-3}$$

For convenience in manipulating the expressions which follow we put

$$R^{-1} = \hat{R} \quad \text{and} \quad Q^{-1} = \hat{Q} \tag{5-4}$$

so that equation (5-3) re-appears in the form

$$\hat{R}(s) = \hat{Q}(s) + H(s) \tag{5-5}$$

Making a perturbation, not necessarily infinitesimal, in $\hat{Q}$, and keeping H(s) fixed, we get the fundamental relationships on which sensitivity studies may be based

$$\delta\hat{R} = \delta\hat{Q} \tag{5-6}$$

where $\delta\hat{R}$ and $\delta\hat{Q}$ denote related changes in $\hat{R}$ and $\hat{Q}$. Let

$$Q' = Q + \delta Q \tag{5-7}$$

and $$\hat{Q}' = \hat{Q} + \delta\hat{Q} \tag{5-8}$$

and define $\hat{Q}'$ to mean

$$\hat{Q}' = (Q')^{-1} \tag{5-9}$$

Note that, in general,

$$\delta\hat{Q} \neq (\delta Q)^{-1} \tag{5-10}$$

Now, from the identity

$$Q'\hat{Q}' = I_m \tag{5-11}$$

we have that

$$Q\delta\hat{Q} + \delta Q\hat{Q} + \delta Q\delta\hat{Q} = 0 \tag{5-12}$$

from which we obtain

$$\delta\hat{Q} = -(Q')^{-1}\ \delta Q\hat{Q} \tag{5-13}$$

An exactly similar set of definitions and identity exists for R' and gives

$$\delta\hat{R} = -(R')^{-1}\ \delta R\hat{R} \tag{5-14}$$

A multivariable feedback system sensitivity matrix is now defined by

$$S' = \left[I_m + Q'H\right]^{-1} \tag{5-15}$$

For vanishingly small perturbations this reduces to

$$S = \left[I_m + QH\right]^{-1} \tag{5-16}$$

which is the inverse of a return-difference matrix.

Thus return-difference matrices play a fundamental role in the sensitivity analysis of multivariable systems.

The relationships derived above may now be used to derive a useful set of system sensitivity properties. Using equations (5-6), (5-13) and (5-14) we have

$$-\hat{R}'\ \delta R\hat{R} = \hat{Q}'\ \delta Q\hat{Q} \tag{5-17}$$

from which

$$\delta R\hat{R} = R'\hat{Q}'\ \delta Q\hat{Q} = \left[Q'(H + \hat{Q}')\right]^{-1}\ \delta Q\hat{Q} \tag{5-18}$$

giving the basic sensitivity formula

$$\delta R\hat{R} = S'\ \delta Q\hat{Q} \tag{5-19}$$

relating a "fractional change" in the closed-loop transfer-function matrix R to the change in the open-loop transfer-function matrix Q. For the special case where the plant matrix G(s) and the controller matrix K(s) are square, suppose we have a perturbation δG in G and that K stays fixed. Then

$$\delta(GK) = \delta GK + G\delta K = \delta GK \tag{5-20}$$

and we find that

$$\delta R\hat{R} = S'\delta G\hat{G} \tag{5-21}$$

and, using the definition of equation (5-9), we get

$$\delta R\hat{R} = S'\left[G'\hat{G} - I_m\right] \tag{5-22}$$

in which the prime must not be dropped for small perturbations. Equation (5-22) is equivalent to a formula derived by Horowitz (Horowitz, 1963) and used by him for synthesis work.

McMorran (McMorran, 1970) has shown that these equations have equivalent complementary forms; for example, equation (5-22) may also be derived in the small-perturbation form

$$\hat{R}\delta R = \hat{Q}\delta Q \left[I_m + HQ\right]^{-1} \tag{5-23}$$

Cruz and Perkins (Cruz and Perkins, 1964) have shown that the matrix S' still retains its significance in the general case when G, K and H are non-square. They have shown that it relates the deviation, e_c say, from the nominal response of the closed-loop system to the deviation, e_o say, of an open-loop system which also contains G and gives the same nominal response, by

$$e_c = S'e_o \tag{5-24}$$

From this they deduce that a sufficient requirement for the closed-loop system to show a reduced deviation is that the matrix

$$M' = \left[(S')^{-1}\right]^{*}(S')^{-1} - I \tag{5-25}$$

must be positive semi-definite for all $s = j\omega$, where * denotes the conjugate transpose. This result is for a general finite perturbation and becomes independent of the change considered in the infinitesimal perturbation case.

6. POLE-SHIFTING AND MODAL CONTROL

It is shown by a simple argument below that, given access to the states of a linear dynamical system, the closed-loop characteristic frequencies may be placed in any desired location. This is a fact of great interest and importance in feedback theory and is the basic idea underlying a great deal of current work. The exploitation of this fact for design purposes led Rosenbrock to propose the modal control technique (Rosenbrock, 1962), which has been investigated by Simon (Simon, 1967), (Simon and Mitter, 1968), Ellis and White (Ellis and White, 1965), Gould (Gould, Murphy and Berkeman, 1970) and others as discussed below.

The transfer-function matrix for a plant having state-space equations of the form of equations (2-21) is

$$G(s) = C(sI - A)^{-1}B \tag{6-1}$$

Suppose, for simplicity of exposition, that the system A-matrix eigenvalues $\lambda_1, \lambda_2, ----, \lambda_n$ are distinct. These are the usual eigenvalues of the matrix A over the complex number field, and should not be confused with the more general eigenvalue concepts (characteristic transfer functions) used elsewhere in this survey. Let U be a matrix whose columns are the system A-matrix eigenvectors, V be a matrix whose rows are the reciprocal eigenvectors of A and Λ be a diagonal matrix of eigenvalues $\lambda_1, \lambda_2, ---, \lambda_n$. Then

$$UV = I_n \tag{6-2}$$

$$UAV = \Lambda \tag{6-3}$$

and it follows that we may express G(s) in the alternative form

$$G(s) = CU\,(sI - \Lambda)^{-1}\,VB \tag{6-4}$$

Consider now the special case in which we have complete freedom over the measurement of system state variables so that we may put, by an appropriate choice of outputs,

$$C = V$$

and thus obtain a transfer function matrix of the form:

$$G(s) = (sI - \Lambda)^{-1}\,VB \tag{6-5}$$

If this transfer function matrix is now used in equations (3-4) and (3-13) we get

$$\det\left\{I_m + (sI - \Lambda)^{-1}\,VBK(s)H(s)\right\} = \frac{\prod_i (s - \gamma_i)}{\prod_j (s - \lambda_j)} \tag{6-6}$$

where γ_i and λ_j denote the set of closed loop and open-loop characteristic frequencies respectively.

Consequently, if we choose controller-matrix and feedback matrix elements so that we have

$$K(s)H(s) = fd^T \tag{6-7}$$

the outer product of an $n \times 1$ matrix f and a $1 \times m$ matrix d^T, then simple determinantal manipulations give (Retallack and MacFarlane, 1970)

$$1 + d^T\,(sI - \Lambda)^{-1}\,BVf = \frac{\prod_i (s - \gamma_i)}{\prod_j (s - \lambda_j)} \tag{6-8}$$

This gives a set of equations from which d^T may be calculated for a specified f to give any desired closed-loop eigenvalue positions. It shows that the resulting loop gains are directly proportional to the eigenvalue shifts imposed. The chief implication of this result is that it shows that arbitrary closed-loop frequency locations can be achieved, given access to all the system states (Wonham, 1967). The technique of altering characteristic frequencies in this way is called modal control. The argument may easily be extended to the case where the system A-matrix has repeated eigenvalues.

The disadvantages of modal control as a design technique are very briefly discussed in Section 11 below. The most complete study of the technique so far as probably that made by Simon (Simon, 1967). The case of incomplete state feedback has been considered (Davison and Goldberg, 1968). Several applications of the technique have been described (Ellis and White, 1965), (Gordon-Clark, 1964),(Davison, 1967).

Porter and Micklethwaite have given procedures for the design of several loops of a multi-loop system either sequentially or simultaneously (Porter and Micklethwaites, 1967). The simultaneous procedures, however, involve the solution of a set of simultaneous non-linear algebraic equations. Porter and Carter have developed a design procedure by which both the real and imaginary parts of any number of pairs of complex conjugate system eigenvalues can be altered in an iterative fashion whereby one complex pair is altered at a time.

7. DUALITY

In feedback control studies two duality concepts are of great value. The first of these is the Kalman duality between feedback control and feedback filtering (Kalman, 1960), (Kalman and Bucy, 1961).

7.1 Kalman Duality Principle

The Kalman Duality Principle (Kalman, 1960), states that for every deterministic feedback control problem there is an equivalent stochastic feedback filtering problem, and vice versa; solution of one problem implies a solution of the dual problem. This duality will be explicitly shown in a later section when stochastic filtering is considered. The physical basis of the duality may be appreciated, however, without at this point involving stochastic considerations by considering the equivalence between the feedback control and feedback filtering operations which is illustrated by figures 7-1 and 7-2.

Figure 7-1 shows a control system in which a controller is used to make the output of an actual system follow the output of an ideal-model system in the face of disturbances to the actual system. The corresponding error-transform vector is

$$e(s) = (I + GK)^{-1} r(s) - (I + GK)^{-1} Gd(s) \qquad (7\text{-}1)$$

Figure 7-2 shows a filter (or signal-recovery) system in which feedback action is used to make the input to a model system follow the input to an actual system in the face of output disturbances to the actual system. The error-transform vector in this case is

$$e(s) = (I + KG)^{-1} r(s) - (I + KG)^{-1} Kd(s) \qquad (7\text{-}2)$$

A comparison of equations (7-1) and (7-2) shows that the two problems have the same mathematical structure and are duals under the appropriate interchanges of inputs and outputs, and the roles of plant and model.

7.2 Dynamical Duality

A dynamical system may be regarded as mapping a set of inputs into a set of outputs. The corresponding dual mapping is obtained by formally interchanging the role of input and output in the mapping (MacFarlane, 1969). A dual dynamical system may then be derived which generates the appropriate dual mapping.

Consider the state-space system of equations:

$$\dot{x} = Ax + Bu \qquad (7\text{-}3)$$

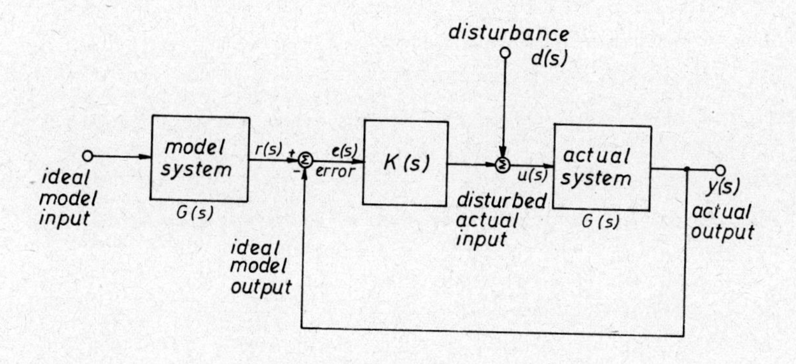

Figure 7-1

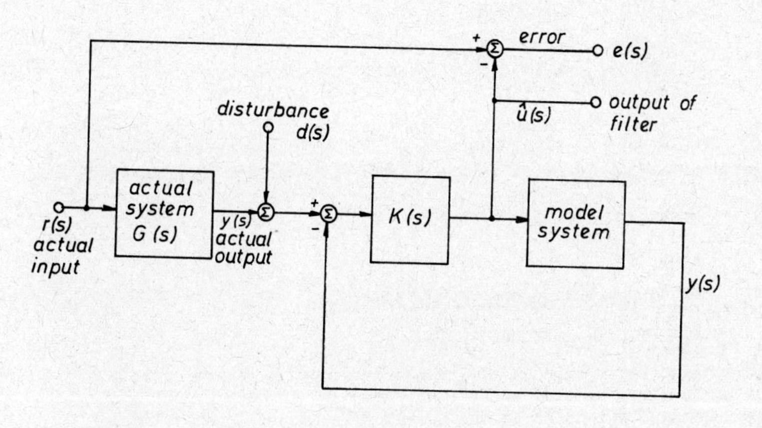

Figure 7-2

$$y = Cx \tag{7-4}$$

The state-transition matrix for equation (7-3) is

$$\emptyset(t,\tau) = \exp\left[A(t - \tau)\right] \tag{7-5}$$

and the corresponding weighting function matrix relating input and output vectors is

$$W(t,\tau) = C\emptyset(t,\tau)\, B \tag{7-6}$$

The input-output mapping is thus given by the matrix convolution integral

$$y(t) = \int_{-\infty}^{t} W(t,\tau)u(\tau)d\tau \tag{7-7}$$

which we may regard as a relationship between "cause" (associated with τ) and "effect" associated with t.

To get the dual dynamical system, we formally interchange the roles of t and τ (i.e. interchange the roles of cause and effect) to get the dual-mapping specified by the convolution integral

$$u(\tau) = \int_{-\infty}^{\tau} W^t(t,\tau)y(t)dt \tag{7-8}$$

$$= \int_{-\infty}^{\tau} B^t \exp\left[-A^t(\tau - t)\right] C^t y(t)dt \tag{7-9}$$

Note the transpositions which arise from the fact that, in the general case, the number of input and output variables may be different. This dual convolution integral is associated with the set of equations

$$\frac{dx^*}{dt} = - A^t x^* + C^t u^* \tag{7-10}$$

$$y^* = B^t x^* \tag{7-11}$$

which define the dual dynamical system.

8. OPTIMALITY AND OPTIMAL CONTROL

Pontryagin (Pontryagin et al., 1963) developed a general theory of optimal control systems using a Hamiltonian approach which is a sophisticated extension of the Hamiltonian method in analytical mechanics (Whittaker, 1927). Kalman (Kalman, 1959), (Kalman, 1963) gave a definitive treatment of the linear case with a quadratic cost function and showed that the appropriate linear feedback controller is determined by the solution of a matrix Riccati equation. In the treatment of the linear optimal feedback controller given here, a simple direct approach to the optimal system equations is outlined using Hamiltonian methods, and a recently developed explicit solution of the matrix Riccati equation is then sketched. The properties of optimal systems are then briefly examined from a frequency-response point of view using the appropriate return-difference matrix properties.

8.1 Hamiltonian Systems

A Hamiltonian system is one governed by a pair of differential equations of the form

$$\dot{x} = \left(\frac{\partial H}{\partial x^*}\right)^t \tag{8-1}$$

$$\dot{x}^* = -\left(\frac{\partial H}{\partial x}\right)^t \tag{8-2}$$

where

$$H = H\ (x, x^*, t)$$

is called the Hamiltonian of the combined system.

These systems have the fundamentally important property that the following stationary principle holds for any solution trajectory (MacFarlane, 1969)

$$\delta \int_o^T L\ dt = 0 \tag{8-3}$$

where L, called the system Lagrangian, is given by

$$L = \langle x^*, \dot{x} \rangle - H \tag{8-4}$$

and 0 and T denote initial and final times.

If a system performance index is chosen as

$$V(T) = \emptyset(T) + \int_o^T L\ dt \tag{8-5}$$

with a free end value of state x, and for fixed initial and final times 0 and T, we have an end-state transversality condition

$$x^*(T) = \left(\frac{\partial V}{\partial x}\right)^t \Bigg|_{t=T} \tag{8-6}$$

which must be satisfied. If the performance index is to be minimized we must have

$$\int_o^T \delta^2 L dt > 0 \tag{8-7}$$

A detailed analysis of the optimal linear feedback system from the Hamiltonian point of view has been given by MacFarlane (MacFarlane, 1969). This shows that if we make the specific choice of Hamiltonian

$$H = \langle x^*, (Ax + BR^{-1}B^t x^*)\rangle - \frac{1}{2} \langle r, Qr\rangle$$
$$+ \langle x, C^t QCr\rangle - \frac{1}{2} \langle x, C^t QCx\rangle - \frac{1}{2} \langle u, Ru\rangle \tag{8-8}$$

equations (8-1) and (8-2) give

$$\dot{x} = Ax + BR^{-1}B^t x^* \tag{8-9}$$

$$\dot{x}^* = C^t QCx - A^t x^* - C^t Qr \tag{8-10}$$

The corresponding system is shown in the block diagram of figure 8-1 and use of equation (8-4) gives the system Langrangian as

$$L = \frac{1}{2} \langle (y - r), Q (y - r)\rangle + \frac{1}{2} \langle u, Ru\rangle \tag{8-11}$$

This shows that the optimal linear servomechanism and regulator problems may be solved by connecting a dynamical system described by the standard state space equation set in feedback cascade with the appropriate dual dynamical system. The Lagrangian shows that departures of the output variable $y(t)$ from a desired value of reference input vector $r(t)$ are penalised by a quadratic-form in the integral of the penalty function $V(t)$ set by the matrix Q, and use of input $u(t)$ is penalised by a quadratic-form in the integral set by matrix R.

The solution of the optimal servomechanism problem is discussed by Sage (Sage, 1968) and MacFarlane(1969). The solution of the optimal regulator problem (that is for $r(t) \equiv 0$) is now considered. Any attempt to implement the solution shown in figure 8-1 will not succeed in practice since the dual dynamical system involved is unstable if the original system is stable. This fundamental attribute of Hamiltonian systems is at the root of all the well-known difficulties of computing numerical solutions to optimal control problems (particularly nonlinear problems) (Bryson and Ho, 1969). In the linear case with quadratic-penalty functions in the performance index integrand, this difficulty was elegantly overcome by Kalman (Kalman, 1959), (Kalman, 1963), and a practicable solution developed. Kalman's technique essentially consists of discarding the unstable dual system and replacing it by an equivalent mapping between state and co-state variables.

8.2 The Matrix Riccati Equation

Suppose we have a system with state-space equations

$$\frac{dx}{dt} = Ax + Bu \qquad y = Cx \tag{8-12}$$

which is to be optimally controlled as a regulator with performance index

$$V(T) = \frac{1}{2} y^t(T)Gy(T) + \frac{1}{2} \int_0^T \left\{ y^t(t)Qy(t) + u^t(t)Ru(t)\right\} dt \tag{8-13}$$

where R is a positive-definite symmetric matrix, and Q and G are non-negative-definite symmetric matrices. Kalman's work showed that the solution can be expressed as

$$u(t) = R^{-1}B^{t}P(\tau)x(t) \tag{8-14}$$

where

$$\tau = T - t \tag{8-15}$$

is the "time-to-go" variable and $P(\tau)$ is the negative-definite solution of the matrix Riccati equation

$$\frac{dP(\tau)}{d\tau} = P(\tau)A + A^{t}P(\tau) + P(\tau)BR^{-1}B^{t}P(\tau) - C^{t}QC \tag{8-16}$$

with boundary condition

$$P(0) = - C^{t}GC \tag{8-17}$$

Until recently the matrix Riccati equation was solved by numerical solution of the corresponding set of first-order differential equations. MacFarlane (MacFarlane, 1963) gave a discussion of the solution in terms of the eigenvectors of the matrix

$$J = \begin{pmatrix} A & BR^{-1}B^{t} \\ C^{t}QC & -A^{t} \end{pmatrix} \tag{8-18}$$

associated with the pair of equations (8-9) and (8-10). Using a similar approach, Potter (Potter, 1966) derived an explicit solution for the asymptotic solution corresponding to an arbitrarily large end-time T. Recently, this line of attack on the problem has been pushed to an ultimate conclusion by Vaughan (Vaughan, 1969) and Walter (Walter, 1970) who have derived explicit solutions for the general linear regulator problem. The solution due to Walter is now sketched (Walter, 1970).

The matrix J given by equation (8-18) has a spectrally symmetrical set of eigenvalues and can be expressed as the product of a set of real matrices obtained from its eigenvalues and eigenvectors:

$$J = \begin{pmatrix} V_{11} & V_{12} \\ V_{21} & V_{22} \end{pmatrix} \begin{pmatrix} \theta & 0 \\ 0 & -\theta \end{pmatrix} \begin{pmatrix} M & 0 \\ 0 & M \end{pmatrix} \begin{pmatrix} V_{22}^{t} & -V_{12}^{t} \\ -V_{21}^{t} & V_{11}^{t} \end{pmatrix} \tag{8-19}$$

In the above expression:

(i) θ is a block diagonal matrix of the form

$$\theta = \begin{pmatrix} \lambda_1 & 0 & - & - & - & - & \cdots \\ 0 & \lambda_2 & & & & & \\ 0 & 0 & \ddots & & & & \\ & & & \sigma_1 & \omega_1 & & \\ & & & -\omega_1 & \sigma_1 & \ddots & \end{pmatrix} \tag{8-20}$$

in which $\lambda_1, \lambda_2, - - - -, \lambda_r$ are the positive-real eigenvalues of J, and the positive-real-part complex eigenvalues of J are $\sigma_1 \pm j\omega_1, - - - -, \sigma_s \pm j\omega_s$ and r + 2s = n, the order of the dynamical system being controlled.

(ii) M is a diagonal matrix of the form

$$M = \begin{pmatrix} 1 & 0 & 0 & - & - & - & \\ 0 & 1 & 0 & - & - & - & \\ & & \ddots & & & & \\ & & 0 & 2 & 0 & & \\ & & 0 & 0 & -2 & \ddots & \\ & & & & & & \end{pmatrix} \qquad (8\text{-}21)$$

where the pairs of elements 2 and -2 are in the positions corresponding to the 2 X 2 blocks of Θ.

(iii) $V_{11}, V_{12}, V_{21}, V_{22}$ are matrices found in the following way. Form a matrix

$$V = \begin{pmatrix} V_{11} & V_{12} \\ V_{21} & V_{22} \end{pmatrix} \qquad (8\text{-}22)$$

by the following procedure. First form a real 2n x 2n matrix X, 2r of whose columns are the real eigenvectors of J in the positions corresponding to the real eigenvalues in the matrices Θ and $-\Theta$. The remaining 2s columns of X are placed in pairs corresponding to the complex eigenvalues in Θ; the first column of each pair is the real part, and the second column is the imaginary part of the eigenvector for the corresponding eigenvalue $\sigma+j\omega$. In a similar way, the remaining 2s columns of the second n columns occur in pairs containing the real and imaginary parts of the eigenvalue for $\sigma-j\omega$ respectively. In order that equation (8-19) is valid, the eigenvectors used in X must be suitably scaled. This is done in the following way:

(a) If

$$X = \begin{pmatrix} V_{11} & X_{12} \\ V_{21} & X_{22} \end{pmatrix} \qquad (8\text{-}23)$$

then we put

$$V = \begin{pmatrix} V_{11} & X_{12} \\ V_{21} & X_{22} \end{pmatrix} \begin{pmatrix} I & 0 \\ 0 & S \end{pmatrix} \qquad (8\text{-}24)$$

where the scaling matrix S is given by

$$S = \left\{ M \left(V_{11}^t X_{22} - V_{21}^t X_{12} \right) \right\}^{-1} \tag{8-25}$$

and S is a block diagonal matrix with 2 x 2 blocks in the same positions as in θ.

Explicit Expressions for Solution of the Optimal Regulator Matrix Riccati Equation. - Two explicit expressions can now be given for $P(\tau)$, one of which contains the steady-state solution $P(\infty)$ as a separate term. The expressions are

$$P(\tau) = \left\{ V_{22} - V_{21} \, Z(\tau) \right\} \left\{ V_{12} - V_{11} \, Z(\tau) \right\}^{-1} \tag{8-26}$$

and

$$P(\tau) = V_{22} \, V_{12}^{-1} + (M V_{12}^t)^{-1} \, Z(\tau) \left\{ V_{12} - V_{11} \, Z(\tau) \right\}^{-1} \tag{8-27}$$

where

$$Z(\tau) = \exp(-\tau\theta) \left\{ M (V_{22}^t + V_{12}^t \, F)(V_{21}^t + V_{11}^t \, F)^{-1} \, M^{-1} \, \exp(-\tau\theta) \right\} \tag{8-28}$$

and

$$F = C^t G C \tag{8-29}$$

Derivations of expressions for the steady-state solution in the form

$$P(\infty) = V_{22} \, V_{12}^{-1} \tag{8-30}$$

were previously given by Potter (Potter, 1966) and MacFarlane (MacFarlane, 1969).

Explicit Expressions for State Trajectories of Optimally Controlled Systems. - Armed with an explicit solution for the matrix Riccati equation we can derive expressions, first given by Walter (Walter, 1970), for the state trajectory of an optimally-controlled linear regulator. Rather more complicated solutions which were given by Vaughan (Vaughan, 1969) require complex matrices and do not recognise the necessity of scaling the eigenvectors of J.

An optimally-controlled trajectory is given by

$$x(t) = \Gamma(t) \, x(o) \tag{8-31}$$

where

$$\Gamma(t) = V_{11} \exp\left\{ (t - T)\theta \right\} W(T) + V_{12} \exp(-t\theta) M \left\{ P(T) \, V_{11} - V_{21} \right\}^t \tag{8-32}$$

and

$$W(T) = -M\left(V_{22}^{t} + V_{12}^{t} F\right)\left(V_{21}^{t} + V_{11}^{t} F\right)^{-1} M^{-1}$$

$$X \exp(-T\theta)\left\{V_{12} - V_{11} Z(T)\right\}^{-1} \qquad (8\text{-}33)$$

An explicit expression for the input $u^{o}(t)$ required for optimal regulation is finally obtained by using the explicit expressions for $P(\zeta)$ and $\Gamma(t)$ in

$$u^{o}(t) = R^{-1}B^{t}\, P(\zeta)\, \Gamma(t)\, x(o) \qquad (8\text{-}34)$$

Frequency Reponse Characterisation of Optimality. - If the performance index to be minimised is

$$V(\infty) = \int_{o}^{\infty} \frac{1}{2}\,(y^{t}Qy + u^{t}Ru)\, dt \qquad (8\text{-}35)$$

then the corresponding optimal control action is given by

$$u = -\,R^{-1}B^{t}Px \qquad (8\text{-}36)$$

where P is the unique positive-definite solution of the steady-state matrix Riccati equation

$$-\,PA - A^{t}P + PBR^{-1}B^{t}P = C^{t}QC \qquad (8\text{-}37)$$

By a sequence of algebraic manipulations (MacFarlane, 1970) this can be converted into an equivalent frequency response form

$$F_{u}^{t}(-s)\, RF_{u}(s) = R + G^{t}(-s)\, QG(s) \qquad (8\text{-}38)$$

where $G(s)$ is the plant transfer-function matrix and $F_{u}(s)$ is the return-difference matrix defined in Section 3.

It may be shown from equation (8-38) (MacFarlane, 1970) that necessary conditions for optimality are that

$$|\rho_{j}(j\omega)| \geqslant 1 \qquad (8\text{-}39)$$

where $\rho_{j}(j\omega)$ are the characteristic loci of $F_{u}(s)$.

It follows directly from this that a necessary condition for optimality is

$$|\det F_{u}(j\omega)| \geqslant 1 \qquad (8\text{-}40)$$

9. FILTERING

The general filtering problem of recovering signals obscured by noise has given rise to a literature at least as large as that on feedback control, so that no attempt will be made to survey it here. The purpose of this section is to briefly outline the way in which feedback enters filtering theory and to give a physical explanation of the "spectral factorization" technique which is important in feedback approaches to filtering and, via duality arguments, to control.

9.1 Stationary Kalman-Bucy Filter

Kalman and Bucy's fundamental work on the optimal filtering problem (Kalman and Bucy, 1961) showed that an optimal filter has a feedback structure and that the stochastic filtering problem and deterministic regulator problem are related by a duality principle. Their work gave a solution for the general time-varying feedback case; in what follows we are only concerned with the stationary case.

Suppose that a random process $x(t)$ is generated by a state-space model

$$\frac{dx}{dt} = Ax + Bu(t) \tag{9-1}$$

and gives rise to an observed signal $z(t)$ corrupted by an additive noise disturbance $v(t)$ so that

$$z(t) = y(t) + v(t) \tag{9-2}$$

where

$$y(t) = Cx(t) \tag{9-3}$$

Let the functions $u(t)$ (which generates a desired message $y(t)$ via the state $x(t)$) and $v(t)$ (which is an unwanted disturbance corrupting the message observation) be independent Gaussian random processes (white noise), with identically zero means and covariance matrices so that

$$\begin{aligned} \text{cov}\,[u(t), u(\tau)] &= Q\delta(t - \tau) \\ \text{cov}\,[v(t), v(\tau)] &= R\delta(t - \tau) \\ \text{cov}\,[u(t), v(\tau)] &= 0 \end{aligned}$$

for all t and τ. Q and R are assumed to be symmetric positive-definite matrices, and $\delta(t - \tau)$ denotes a Dirac delta function. The stationary Kalman-Bucy filter for generating a minimum-variance linear unbiased estimate $\hat{y}(t)$ of $y(t)$ is shown in figure 9-1. The filter has a feedback structure with a gain K given by

$$K = PC^TR^{-1} \tag{9-4}$$

where P satisfies the steady-state matrix Riccati equation (Kalman and Bucy, 1961)

$$-PA^T - AP + PC^TR^{-1}CP = BQB^T \tag{9-5}$$

and is the convariance matrix for the error in the state estimate $(\hat{x} - x)$

9.2 Frequency Response Form of Optimal Filtering Equation

MacFarlane has shown (MacFarlane, 1971) how the steady-state matrix Riccati equation may be expressed in frequency response form as

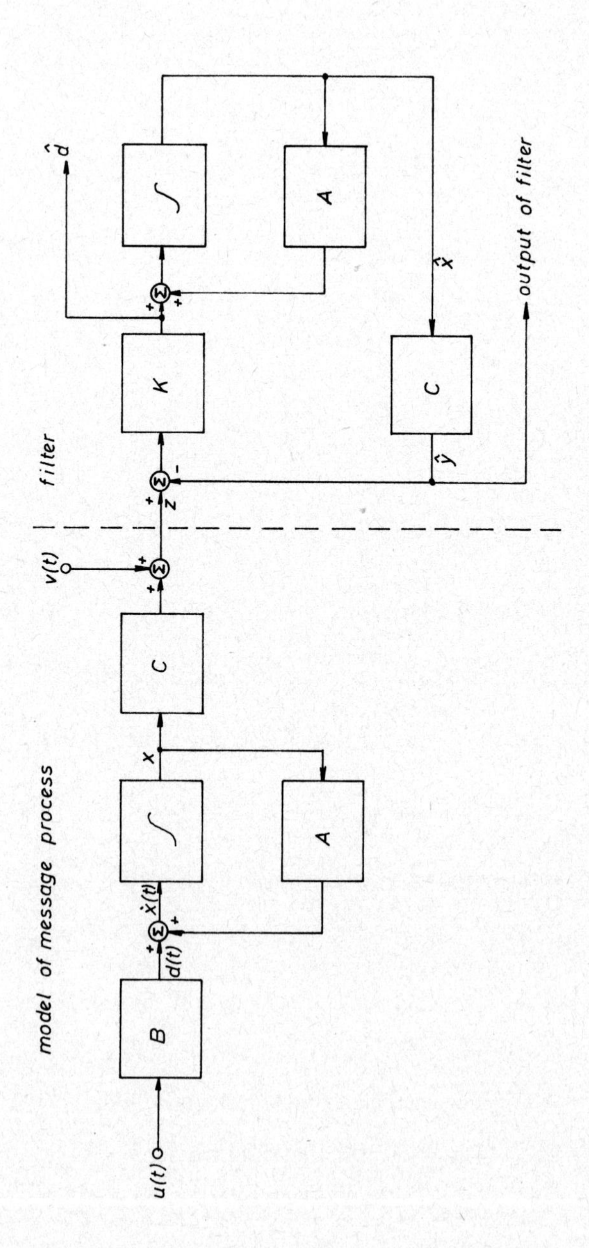
model of message process
filter
u(t)
B
d(t)
ẋ(t)
A
x
C
v(t)
z
K
d̂
A
x̂
C
ŷ
output of filter

Figure 9-1

$$F(s)RF^t(-s) = R + G(s)QG^t(-s) \tag{9-6}$$

where

$$F(s) = I + C(sI - A)^{-1}K \tag{9-7}$$

9.3 Feedback Interpretation of Spectral Factorization

Equation 9-6 has a most interesting form, and gives an illuminating physical insight into the working of a stationary optimal filter. To see the significance of the matrix F(s), consider the feedback-filter arrangement shown in figure 9-2. Suppose that all the feedback loops are broken, as shown in figure 9-2, and a (deterministic) signal-transform vector $\alpha(s)$ is injected at point a. The transform of the signal vector returns at a' is then

$$-C(sI - A)^{-1}K\alpha(s)$$

and the difference between the injected and returned signal-transform vectors is thus

$$\left\{I + C(sI - A)^{-1}K\right\}\alpha(s) = F(s)\alpha(s) \tag{9-7}$$

where F(s) is defined as the 'return-difference matrix' for the feedback system. Note that the loops have been broken at a different point from that in the controller case; this is for duality reasons, to keep the final equations of both filter and controller of the same form.

With all the feedback loops closed, as shown in figure 9-3, we find that the 'tracking-error' transform vector e(s) satisfies the equation

$$e(s) = z(s) - C(sI - A)^{-1}Ke(s) \tag{9-8}$$

from which we have

$$\left\{I + C(sI - A)^{-1}K\right\}e(s) = z(s)$$

so that

$$e(s) = F^{-1}(s)z(s) \tag{9-9}$$

Suppose that the spectral-density matrix of the observation z is spectrally factorised as

$$R + G(s)QG^T(-s) = \Delta(s)\Delta^T(-s) \tag{9-10}$$

where $\Delta(s)$ is the Hurwitz factor of the spectral factorization (having all poles and zeros in the left-half complex plane). Then

$$F(s)R^{1/2} = \Delta(s)$$
$$F(s) = \Delta(s)R^{-1/2} \tag{9-11}$$

$$F^{-1}(s) = R^{1/2}\Delta^{-1}(s) \tag{9-12}$$

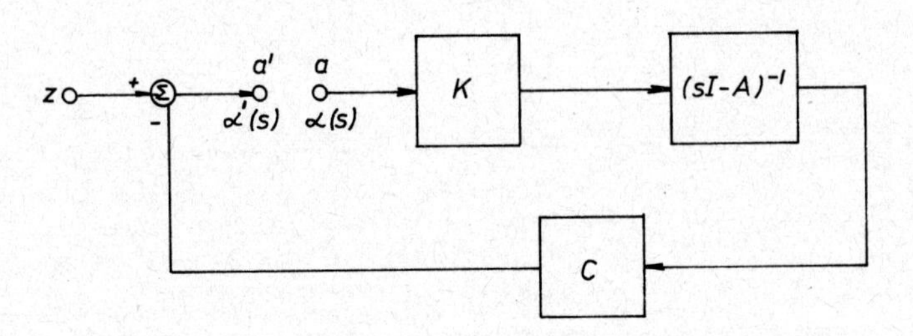

Figure 9-2

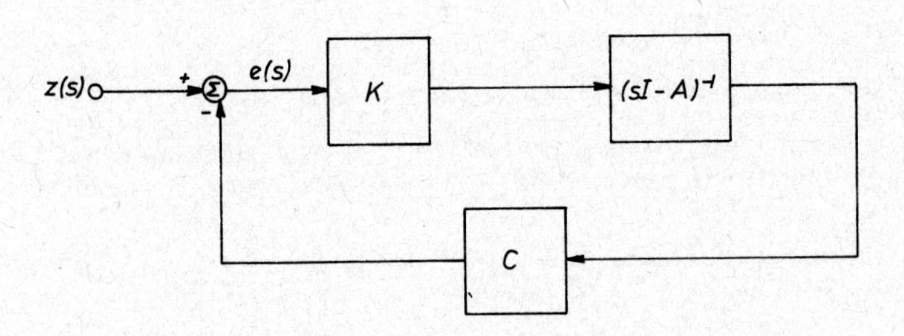

Figure 9-3

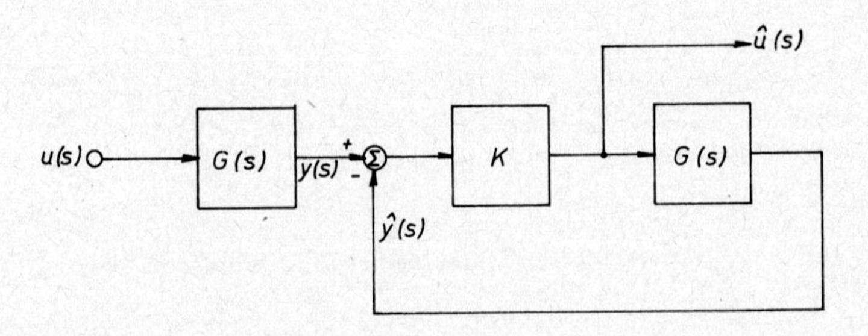

Figure 9-4

We therefore deduce that the spectral factorization of the spectral-density matrix of the observation simply generates the return-difference matrix of the optimal feedback filter. Furthermore, the tracking-error transform vector is given by

$$e(s) = R^{1/2}\Delta^{-1}(s)z(s) \tag{9-13}$$

and it follows from the fact that $\Delta^{-1}(s)$ is the 'whitening filter' for the observations that the tracking error is generated by feeding white noise with a unit covariance matrix through a non-dynamic transference $R^{1/2}$.

The essence of the way in which feedback is used to recover a signal originating behind known dynamics may be seen by considering the simple deterministic system shown in figure 9-4. Suppose that u(s) is a vector signal, not directly accessible for measurement, but originating behind a known transfer-function matrix G(s). A copy of G(s) is built, and has a multivariable feedback system closed round it through a controller matrix K. If we can design K so that high loop gains can be achieved over a desired operating-frequency range, $\hat{y}(s)$ will closely approximate to y(s), and thus $\hat{u}(s)$ will closely approximate to the desired signal u(s). The feedback filter of figure 9-1 essentially works in this way, with the additional sophistication of a return-difference adjusted to strike the best compromise between signal and noise. In figure 9-1, the controller K drives a copy of the state-transition mechanism of the message-generating process, and the return-difference matrix F(s) of the feedback filter is adjusted so that

$$F(s)RF^{T}(-s) = R + G(s)QG^{T}(-s) \tag{9-14}$$

which shows immediately that the loop gains are increased when the signal/noise ratios are high, and reduced when they are low. The frequency-response version of the steady-state matrix Riccati equation thus has the following three physical implications:

(a) It shows how an optimal filter adjusts, for all values of s, the loop gains round a copy of the message state-transition process to obtain the best compromise between signal and noise vectors at any given value of s.

(b) Since it virtually constitutes a spectral factorization of the spectral-density matrix of the observations being processed, it shows how the return-difference matrix is simply relared to the 'whitening filter' for the observations. This, in turn, shows directly how the tracking error is a white-noise process which has the same covariance matrix as the corrupting noise process.

(c) It shows how the optimal filter and optimal regulator are virtually the same problem from the feedback-control point of view. In an optimal feedback filter, the loop gains are adjusted at all frequencies to effect the best compromise between signal- and noise-power spectral densities; in an optimal regulator, the loop gains are adjusted to effect the best compromise between input cost penalties and output cost penalties.

The problem of spectral factorization has been extensively studied. Youla gave a definitive study (Youla, 1961) in which he derived a method for the factorization of rational continuous-time spectral density matrices, proving his main theory by using the Smith-McMillan form of the canonical matrices (McMillan, 1952). Davis produced a method (Davis, 1963) which avoids the use of Smith-McMillan forms by using a sequence of transforming matrices which, suitably applied, transform a given spectral density matrix into a unit matrix. Csaki and Fisher then extended Davis's ideas to give an exact method of spectral factorization (Csaki and Fisher, 1967). Kavanagh further modified the Davis method to give a reduction to a matrix having second degree rather than a constant, which gives a useful saving of computational effort (Kavanagh, 1968).

10. OBSERVERS

The previous discussions of optimal control and pole-shifting have shown how state variables play a key role in certain time-domain approaches to multivariable feedback control. The simplest physical explanation for the success of these techniques, when viewed from the frequency-response point of view, is that accessing states generates a mechanism for providing phase-advance in the system closed loops, and thus ensures adequate stability margins in the feedback design.
If the systems states are not accessible to measurement, then severe penalties are incurred in terms of phase lag. Techniques have therefore been developed, based principally on the work of Luenberger (Luenberger, 1964), (Luenberger, 1966), to provide estimates of inaccessible states. This leads to the feedback control configuration shown in figure 10-1 where the Observer provides the Controller with an estimate $\hat{z}$ of a suitable linear combination of state variables.

Suppose z is a vector whose elements are linear combinations of the system states so that

$$z = Lx \tag{10-1}$$

Let $\hat{z}$ be an estimate of z, and Δz the associated error in the estimate then

$$\hat{z} - z = \Delta z \tag{10-2}$$

Now the Observer, as shown by figure 10-1, is a linear dynamical system driven by y and u, the controlled system output and input respectively. Suppose it satisfies the equations

$$\frac{d\hat{z}}{dt} = D\hat{z} + TCx + Eu \tag{10-3}$$

where D, T and E are constant matrices. A calculation now shows that

$$\frac{d}{dt}(\Delta z) = D(\Delta z) + (DL + TC - LA)x + (E - LB)\,u \tag{10-4}$$

Therefore, if we choose the matrices D, T and E (which define the Observer construction) so that

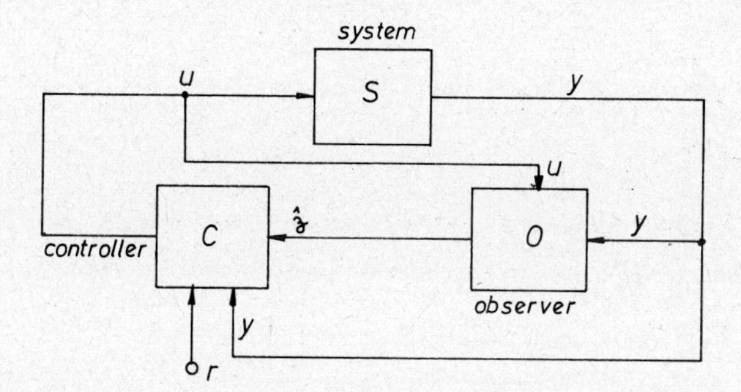

Figure 10-1

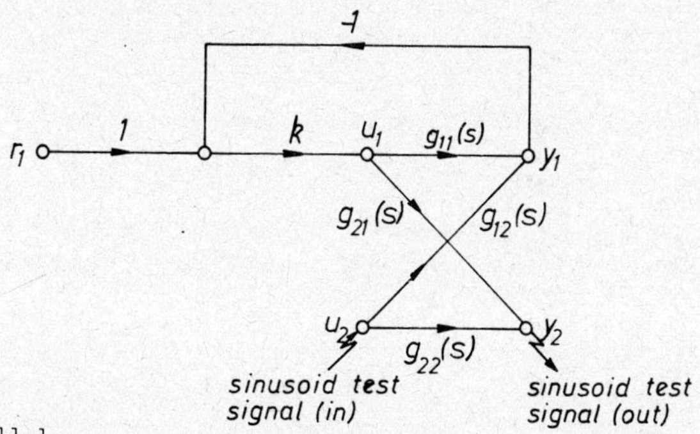

Figure 11-1

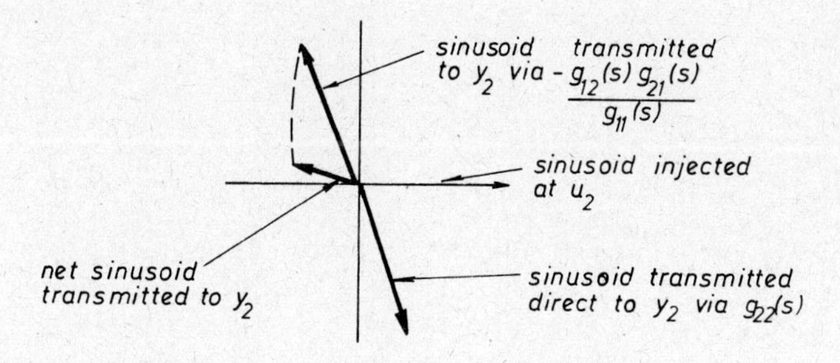

Figure 11-2

$$DL + TC = LA \tag{10-5}$$

$$\text{and} \quad E = LB \tag{10-6}$$

then the dynamics of the error in the estimates are given by

$$\frac{d}{dt}(\Delta z) = D(\Delta z) \tag{10-7}$$

If D is a stable matrix, then Δz will decay exponentially with time and

$$z \to Lx \quad \text{as} \quad t \to \infty$$

independently of the plant input u, since equations (10-6) shows that Δz is uncontrollable from u. Because of this uncontrollability, the Observer simply adds its own (arbitrarily specified) eigenvalues to those of any closed-loop system designed as if the state vector were available for feedback purposes. This 'asymptotic tracking' feature of the Observer is most valuable; after the decay of observer-error transients (caused say by disturbances to the plant but not the Observer), the Observer "follows" the plant states with a dynamic tracking error dependent on the choice of the Observer system eigenvalues.

Supposes a certain proportional matrix-feedback control law is required so that

$$u = Kx \tag{10-8}$$

and that this is expressed in terms of the available output y and the estimates z as

$$\begin{aligned} u &= K_1 y + K_2 z \\ &= K_1 Cx + K_2 Lx \end{aligned} \tag{10-9}$$

so that

$$K = K_1 C + K_2 L \tag{10-10}$$

If we apply an estimated feedback control action by putting

$$\hat{u} = K_1 y + K_2 \hat{z} \tag{10-11}$$

then it is found that the closed-loop dynamics of the combined system plus the Observer are

$$\frac{d}{dt}\begin{pmatrix} x \\ \Delta z \end{pmatrix} = \begin{pmatrix} (A+BK) & BK_2 \\ 0 & D \end{pmatrix}\begin{pmatrix} x \\ \Delta z \end{pmatrix} \tag{10-12}$$

This is an extremely interesting and valuable result and consequently great interest has been aroused in the design of such deterministic estimators. It shows that the poles of the closed-loop dynamic system are as they would have been had the originally-desired state-feedback been directly implemented.

Retallack (Retallack, 1970) has given a study of observer design from the transfer-function matrix point of view.

In a discussion of multivariable observer design, Luenberger indicated that the procedure may be simplified by making use of his techniques for single-output plants, together with a canonical form which uses a linear co-ordinate transformation to put the plant equations into the form of m component sub-systems each with a single output. (Luenberger, 1966). Simon discussed observer design as the dual problem to the arbitrary location of closed-loop system eigenvalues with state feedback (Simon, 1967). Porter and Woodhead discussed the effect of a controller (Porter and Woodhead, 1968). This topic was also considered by Bongiorno and Youla. They studied (Bongiorno and Youla, 1968) the use of observers with control schemes where the steady-state solution of an appropriate matrix Riccati equation is used with an estimated state feedback to minimise a quadratic-form performance index. Their work showed that the increase in performance cost due to observer errors does not become arbitrarily small as the modulus of the observer eigenvalues becomes arbitrarily large. Furthermore, no constant feedback law other than that given by the matrix Riccati equation will reduce this additional cost for all initial plant states, if the system is to be stable.

11. SURVEY OF DESIGN TECHNIQUES

Before giving an account of some of the currently-available multivariable feedback design techniques, it will be useful and appropriate to consider certain general aspects of the design problem. One key question must be disposed of straight away: why not simply design a succession of feedback loops one at a time using well-established single-input single-output feedback theory? This question is not a naive one and certainly cannot be answered by merely pointing to the power and elegance of state-feedback theory. To shed some light on this, we will consider the specific example of a system with two manipulated inputs and two controlled outputs.

The difficulties in such a multiple-loop feedback control system arise from interactive effects; the action of one feedback loops affects the action of the other. This interaction is normally such that the stability margins of system operation are reduced, and it is essential to design multiple-loop systems in such a way as to avoid interactive effects which are prejudicial to system stability. It is helpful to have some physical insight into the nature of these difficulties before embarking on any detailed analytical study. For this purpose, the nature of the difficulties are considered for the problem of using two feedback loops to control a plant with the transfer-function matrix

$$G(s) = \begin{pmatrix} g_{11}(s) & g_{12}(s) \\ g_{21}(s) & g_{22}(s) \end{pmatrix} \tag{11-1}$$

Suppose that we set up the two loops to control output 1 by feedback to input 1, and output 2 by feedback to input 2. To see how difficulties of control arise, consider the signal transmission between input 2 and output 2 with the top feedback loops closed, as shown in figure 11-1. A simple calculation gives

$$y_2(s) = \left[g_{22}(s) - g_{12}(s)k_1\left\{1 + k_1 g_{11}(s)\right\}^{-1} g_{21}(s)\right] u_2(s) \qquad (11\text{-}2)$$

If the top feedback-loop gain is high, so that

$$|k_1 g_{11}(s)| \gg 1 \qquad (11\text{-}3)$$

for all s, then

$$y_2 \simeq \left\{g_{22}(s) - \frac{g_{12}(s)g_{21}(s)}{g_{11}(s)}\right\} u_2(s) \qquad (11\text{-}4)$$

Thus, for sufficiently high gains in the top feedback loop, the net transmission between input 2 and output 2 is

$$\left\{g_{22}(s) - \frac{g_{12}(s)g_{21}(s)}{g_{11}(s)}\right\} \qquad (11\text{-}5)$$

with corresponding frequency-response characteristic

$$\left\{g_{22}(j\omega) - \frac{g_{12}(j\omega)g_{21}(j\omega)}{g_{22}(j\omega)}\right\} \qquad (11\text{-}6)$$

Difficulties of control arise from the fact that, in addition to the direct transference $g_{22}(j\omega)$ between input 2 and output 2, the action of the top controller causes an extra transmission term

$$\left\{-g_{12}(j\omega)g_{21}(j\omega)/g_{22}(j\omega)\right\}$$

to appear.

As a specific example, suppose that $g_{11}(s)$, $g_{12}(s)$, $g_{21}(s)$ and $g_{22}(s)$ are all 1st-order lags; no single loop control difficulties can then arise. The phase relationship implied by figure 11-6, however, may be represented by the conventional phasor diagram of figure 11-2. The action of the top controller may then be such as to add a large amount of additional phase lag between input 2 and output 2, the additional signal lagging on the directly transmitted signal by the order of an additional 180°. The presence of the top controller injecting extra phase lag into the bottom transference will obviously greatly reduce the amount of loop gain which can be applied in the feedback loop of the bottom controller without causing instability. The stable operating region, as a function of both loop gains may then have the form shown in figure 11-3. This shows that the system may then go unstable if high gains are applied in both loops, although it may tolerate an arbitrarily high gain on either loop alone.

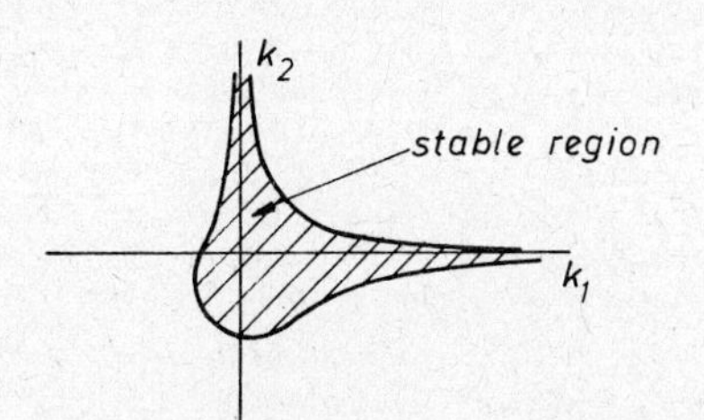

Figure 11-3

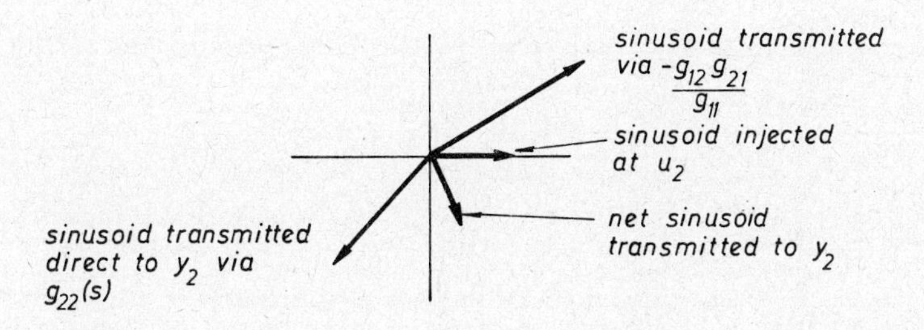

Figure 11-4

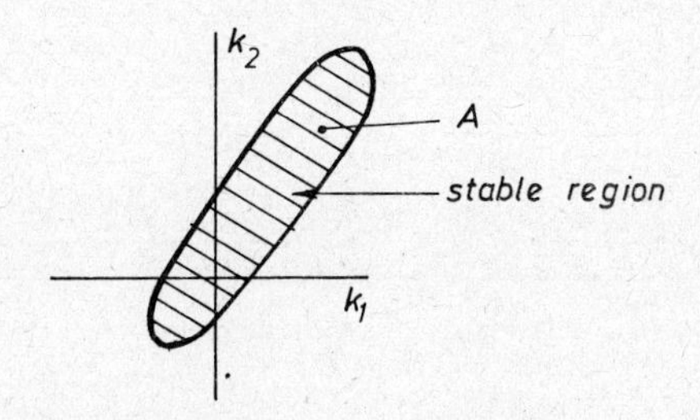

Figure 11-5

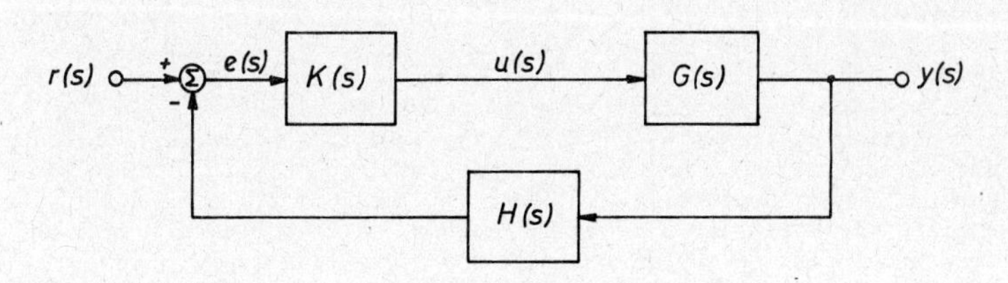

Figure 11-6

As a further example, showing the interaction operating in the other sense, suppose that the phasor display of direct and injected sinusoids for some other specific system is as shown in figure 11-5. Practical difficulties still arise, since any attempt to operate the system at point A with high gains on both loops will result in instability, should either output variable transducer fail, and thus the system will have a low integrity.

With the bottom loop closed and the top loop open, the net transmission between input 1 and output 1 is given by the transfer function

$$\left[g_{11}(s) - g_{21}(s)k_2\left\{1 + k_2 g_{22}(s)\right\}^{-1} g_{12}(s)\right] \tag{11-7}$$

and the same arguments apply. Expr. (11-5) can be written as

$$\frac{g_{11}(s)g_{22}(s) - g_{12}(s)g_{21}(s)}{g_{11}(s)} = \frac{\det G(s)}{g_{11}(s)} \tag{11-8}$$

and, if det G(s) has any zeros in the right-half complex plane, the transmission between input 2 and output 2 with the top feedback loop closed will become nonminimum phase, and control difficulties will become acute. This shows the situation in an extreme form, since it is well known that nonminimum phase transference causes difficulties in feedback control schemes.

Further study of more general examples, particularly those in which det G(s) vanishes in the right-half plane, will show that these effects persist, regardless of the particular arrangements of inputs and outputs between which the two feedback loops are connected. They may only be avoided by using the general feedback scheme shown in figure 11-6 with off-diagonal elements present in the controller matrix K(s). Most of the difficulties in multiple-loop feedback control stem from the necessity to design for ALL the entries in K(s).

We conclude that only two systematic procedures are suitable for multiple-loop control-system design:

(a) to design all the loops together
(b) to design the loops in a sequential manner, taking account of all interactions as the process unfolds.

State-space methods and algebraic synthesis techniques can be thought of as "all-the-loops-together" design approaches. Experience of certain kinds of industrial design problems seems to show that such methods lack the flexibility needed to handle certain kinds of engineering constraint such as (i) that the controller must have high integrity, (ii) must not be too complicated and (iii) must not use too much phase advance in any loop. The need to try and handle engineering constraints of this sort has led to the development of frequency-response design methods of a sequential type.

The characteristic transfer-function concepts introduced in Section 3 can be used to discuss certain general features of any feedback design for the basic configuration shown in figure 11-6, so these are examined next.

Let $G(s)K(s) = Q(s)$ (11-9)

and note that Q(s) will always be square since r(s) and y(s) will be of the same order.

Let the eigenvalues of the matrix Q(s), defined in equation (3-19), be $q_1(s)$, $q_2(s)$, . . ., $q_m(s)$, and suppose that they are distinct in the extension field. (If they are not distinct, an arbitrarily small set of perturbations in the elements of Q(s) will make this so). Let a corresponding set of eigenvectors be $w_1(s)$, $w_2(s)$, . . ., $w_m(s)$. From standard algebraic theory, this vector set will constitute a basis for a linear vector space of dimension m over the field of rational functions. Define a matrix W(s) by

$$W(s) = \left[w_1(s)w_2(s) \; . \; . \; . \; w_m(s)\right] \qquad (11\text{-}10)$$

Since the vectors $w_1(s)$ through $w_m(s)$ are linearly independent, W(s) is invertible, and we have

$$W^{-1}(s)Q(s)W(s) = \text{diag}\left[q_1(s)q_2(s) \; . \; . \; . \; q_m(s)\right]$$
$$= \Lambda^Q(s) \text{ say} \qquad (11\text{-}11)$$

Let

$$W^{-1}(s) = V(s) = \begin{pmatrix} v_1^T(s) \\ v_2^T(s) \\ \cdot \\ \cdot \\ \cdot \\ v_m^T(s) \end{pmatrix} \qquad (11\text{-}12)$$

where $v_1^T(s)$, . . ., $v_m^T(s)$ are the rows of V(s)

Then $V(s)Q(s)W(s) = \Lambda^Q(s)$ (11-13)

$W(s)\Lambda^Q(s)V(s) = Q(s)$ (11-14)

and the system closed-loop transfer-function matrix R(s) may be written in the form

$$R(s) = \left\{I_m + W(s)\Lambda^Q(s)V(s)H(s)\right\}^{-1}W(s)\Lambda^Q(s)V(s) \qquad (11\text{-}15)$$

and we have

$$R^{-1}(s) = W(s)\left\{\Lambda^Q(s)\right\}^{-1}V(s) + H(s) \qquad (11\text{-}16)$$

and, putting $s = j\omega$ to discuss frequency response,

$$R^{-1}(j\omega) = W(j\omega)\left\{\Lambda^Q(j\omega)\right\}^{-1}V(j\omega) + H(j\omega) \qquad (11\text{-}17)$$

Suppose that, at some specific ω_e, all the eigenvalues of $Q(j\omega)$ have arbitrarily large modulus. Then all the elements of $\Lambda^Q(j\omega_e)$ will be arbitrarily large, and we shall have

$$R^{-1}(j\omega_e) \simeq H(j\omega_e)$$

$$R(j\omega_e) \simeq H^{-1}(j\omega_e)$$

This leads to the important conclusion that arbitrarily good control action will be exercised over any frequency range over which the moduli of all the eigenvalues of $Q(j\omega)$ are arbitrarily large.

Conversely, suppose that, at some other specific frequency ω_h, all the eigenvalues of $Q(j\omega)$ have arbitrarily small modulus. Then all the elements of $\mathcal{R}(j\omega_h)$ will be arbitrarily small and we will have

$$R^{-1}(j\omega_h) \simeq Q^{-1}(j\omega_h)$$

$$R(j\omega_h) \simeq Q(j\omega_h) \tag{11-18}$$

For any form of design procedure to achieve a multivariable feedback control, two sets of conditions must be satisfied by the eigenvalues of $Q(j\omega)$:

(a) The moduli of all the eigenvalues of $Q(j\omega)$ must be large over any frequency range for which it is desired to exercise good control. Specifically, all the eigenvalues of $Q(0)$ must be large if good control action is required at d.c. and very low frequencies.

(b) To satisfy the closed-loop-stability criteria of Nyquist type imposed on the characteristic frequency-response loci, all the eigenvalues of $Q(j\omega)$ must have small moduli at high frequencies.

Now consider the response, under multiple-loop feedback control, of a system satisfying both these conditions to a set of step changes in the reference inputs. Thus, let the reference input-transform vector be

$$r(s) = \frac{1}{s}\bar{r} \tag{11-19}$$

where r is a vector whose elements are real constants. The above arguments on the behaviour of $R(j\omega)$ at low and high frequencies, together with use of the final-value and initial-value theorems of Laplace transform theory, then enable us to infer the initial and final values of system-controlled outputs under feedback control as

$$y(0+) = Q(\infty)\bar{r} \tag{11-20}$$

$$y(\infty) = H^{-1}(0)\bar{r} \tag{11-21}$$

Commutative Controller. - The general problem of multivariable-feedback-control-system design can now be looked on in terms of choosing the controller matrix $K(s)$ so that the eigenvalues of the matrix $G(s)K(s)$ have certain prescribed properties. This is the viewpoint taken in the general discussion of design techniques given in this Section. The crux of the design is choosing the individual elements of $K(j\omega)$ to achieve the desired modification of the characteristic loci of $G(j\omega)$; this is difficult since little is known about the location of the eigenvalues of the product of two matrices in terms of the eigenvalues of the individual matrices which are multiplied together. One particular situation may be handled, however, in a simple way; namely that in which the two matrices commute, and therefore have a common set of eigenvectors. This leads to the theoretical concept of what may be called a commutative controller. A commutative controller would be designed by sythesising m clasical single-loop controllers in the eigenframework of the plant transfer-function matrix, and then 'transforming back' to the original reference basis to get the required controller matrices. (MacFarlane, 1970).

Let $\gamma_1(s), \ldots, \gamma_m(s)$ be the eigenvalues (characteristic transfer functions) of the plant transfer-function matrix G(s), with a corresponding eigenvector set $w_1(s), \ldots, w_m(s)$, and associated matrices W(s), V(s) as defined by equations (11-10) and (11-12). For a commutative controller, the eigenframework of G(s) and Q(s) will be the same. G(s) can then be expressed in dyadic form as

$$G(s) = \sum_{j=1}^{m} w_j(s)\gamma_j(s)v_j^t(s) \tag{11-19}$$

Let $$K(s) = \sum_{k=1}^{m} w_k(s)k_k(s)v_k^t(s) \tag{11-20}$$

and $$H(s) = \sum_{i=1}^{m} w_i(s)h_i(s)v_i^t(s) \tag{11-21}$$

Then $$G(s)K(s) = \sum_{j=1}^{m} w_j(s)\gamma_j(s)k_j(s)v_i^t(s) \tag{11-22}$$

since

$$v_i^t(s)w_j(s) = \begin{cases} 1 \text{ if } i = j \\ 0 \text{ if } i \neq j \end{cases} \tag{11-23}$$

as we have

$$V(s)W(s) = I_m \tag{11-24}$$

Simple manipulations then give

$$R(s) = \{I_m + G(s)K(s)H(s)\}^{-1}G(s)K(s)$$

$$= \sum_{j=1}^{m} w_j(s)\left\{\frac{\gamma_j(s)k_j(s)}{1+\gamma_j(s)k_j(s)h_j(s)}\right\}v_j^t(s)$$

$$= W(s)\Lambda(s)V(s) \tag{11-25}$$

where

$$\Lambda(s) = \text{diag}\left\{\frac{\gamma_j(s)k_j(s)}{1+\gamma_j(s)k_j(s)h_j(s)}\right\} \tag{11-26}$$

and $K(s) = W(s)\{\text{diag } k_i(s)\}V(s)$ (11-27)

$$H(s) = W(s)\{\text{diag } h_i(s)\}V(s) \quad (11\text{-}28)$$

From eqn. (11-25), it follows that

$$R(s) \to \sum_{j=1}^{m} w_j v_j^t = I_m \text{ as } k_j \to \infty, j = 1, \ldots, m$$

if $h_j(s) = 1$ and $j = 1,2, \ldots, m$.

The above relationships exhibit the idea of the commutative controller. In simple physical terms, the transformation of basis to the plant eigenframework means expressing general signal-transform vectors as linear combinations of those vectors (the eigenvectors of G(s)) to which the plant appears as a set of simple single scalar transfer functions, namely the characteristic transfer functions or eigenvalues $\gamma_j(s)$. In this basis, we would try to carry out m single-loop designs to choose the set of single-loop controllers $k_j(s)$ and feedback operators $h_j(s)$, and finally transform back to obtain the actual controller K(s) and feedback operator H(s) by means of eqns. (11-27) and (11-28).

Direct Synthesis. - As explained below, the commutative controller approach is not a practicable procedure in most cases of controller design. Other ways, particularly the characteristic locus method outlined below, have been found to exploit eigenvalue-loci for feedback design purposes. It is interesting however, before leaving the general features of the problem, to derive a simple but instructive formula for "direct synthesis" of controller action in the special case where the plant matrix G(s) is square.

Since

$$Q = W\Lambda^Q W^{-1} = GK \quad (11\text{-}29)$$

we may put, when G(s) is not identically singular,

$$K(s) = G^{-1}(s)W(s)\Lambda^Q(s)W^{-1}(s) \quad (11\text{-}30)$$

Such a formula could, in principle, be used for a direct synthesis of multivariable feedback controllers for simple systems, choosing W and Λ^Q to achieve a required closed-loop performance. This "brute-force" synthesis is unlikely to work for a complicated practical problem, but nevertheless the form of equation (11-30) is very revealing and allows us to make an important general observation. Since

$$G^{-1}(s) = \frac{\text{adjoint } G(s)}{\text{determinant } G(s)} \quad (11\text{-}31)$$

we see, in a more general context than the simple discussion given above, how right-half plane zeros of det G(s) can cause great difficulty in multivariable feedback design. In formula (11-30) W and Λ set the properties of the closed-loop system and must therefore satisfy normal realizeability constraints and be such that the diagonal elements of Λ

satisfy suitable Nyquist stability requirements. Thus any right-half plane zeros of det G(s) raise serious stability problems in addition to the realizeability difficulties normally associated with approximating to an inverse transfer function.

We now have available all the material required to give a unified and systematic outline of available design techniques. The stability and optimality characteristics of the characteristic frequency response loci of Q(s) serve in this as a link between classical single-loop frequency-response methods and multivariable-system design techniques.

11.1 Non-interacting Control Technique

If we choose K(s) so that G(s)K(s) is a diagonal matrix, the resulting system is said to be non-interacting, and we can then obviously complete a design by using single-loop frequency response design techniques if H(s) is also diagonal. This approach to design was first put forward by Boksenbom and Hood (Boksenbom and Hood, 1949). The design problem then reduces to the design of m "uncoupled" single-loop controllers whose frequency responses are, trivially, the system characteristic loci. If this technique were of wide applicability then the general problem of designing multivariable feedback controllers would virtually cease to exist. Even for fairly simple situations, however, it has several severe disadvantages which are as follows:

(i) Too much of the limited design freedom available for choosing the elements of K(s) is used up in making G(s)K(s) diagonal, leaving little room for manoeuvre in compensating the diagonal transferences obtained. This usually results in poor dynamic performance of the resulting closed-loop system.

(ii) Even if the technique gives an acceptable control performance, the form of the resulting controller is usually unnecessarily complicated.

(iii) As discussed by Rosenbrock (Rosenbrock, 1966), the procedure suffers a crippling disadvantage when det G(s) has any right-half plane zeros, the technique then giving poor or unstable control.

The non-interacting design technique has been discussed by Tsien (Tsein, 1954) and Rae (Rae, 1964).

11.2 Modal-control Technique

Section 6 has shown how, given access to all the states of a system described in state-space terms, a linear proportional-feedback controller may be built up which places the system closed-loop state-space eigenvalues in any desired position. The disadvantages of the use of this algorithm as a design technique are:

(a) It leads to a simple proportional controller, and gives no guidance as to the choice of dynamic compensating elements.

(b) No means are provided for correlating closed-loop transient response (which involves the zeros of all the transfer functions concerned) with the positions achieved for the poles of the closed-loop transfer functions.

(c) Linear combinations of system states (the system modes) are controlled, and not the states themselves. In certain circumstances, this can lead to a situation where the modally controlled system has poor disturbance-rejection properties, and relatively large disturbances of state occur even with high loop gains in the controller.

11.3 Optimal-control Technique

The discussion of Section 8 has shown that the characteristic frequency-response loci of an optimal proportional-feedback controller have infinite gain margin and at least 60° phase margin. This follows from simple geometrical consequences of the fact that the characteristic loci do not penetrate the unit disc surrounding the critical point in the complex plane. The disadvantages of the optimal-control approach are therefore:

(i) It requires access to all the system states.

(ii) It provides gain margins far in excess of these actually required for stability.

(iii) It offers no means of providing dynamic compensation.

Reflection on the mechanism of operation of an optimal controller in terms of its characteristic frequency-response loci will show that the accession of all system states is what gives the degree of phase advance required to achieve the arbitrarily high gain margins of the optimal characteristic frequency-response loci.

11.4 Commutative-controller Technique

The commutative-controller technique is at first sight a conceptually appealing one, since it operates directly on the characteristic loci which control the feedback system behaviour. It has, however, several severe disadvantages:

(a) The characteristic transfer functions of G(s) will normally be irrational. This leads to sever computational difficulties which are only partially alleviated by approximation of the irrational quantities by rational functions in s.

(b) At high frequencies, the stability considerations become paramount, and the characteristic loci moduli must be made small. As shown in the discussion leading up to eqn. (11-20), the closed-loop response immediately after a transient input is largely determined by Q(s). This means that the transient response of the closed-loop system cannot be completely controlled by designing a set of m scalar systems.

If Q(s) has significant off-diagonal terms at high frequencies, interaction terms cannot be suppressed by the design of m scalar systems which specify the characteristic loci behaviour. This is simply because the stability requirements reduce high-frequency gain to the point at which high-frequency cross-couplings cannot be suppressed. The only straightforward way to eliminate such interaction is to make Q(s) of diagonally-dominated form, as discussed in the Inverse Nyquist

Array technique below. The unique properties of the characteristic loci can be exploited in another way for design purposes as discussed in the Characteristic Locus Method below.

11.5 Inverse Nyquist Array Technique

The Inverse Nyquist Array technique (Rosenbrock, 1969) introduced by Rosenbrock is a very valuable design technique for a wide range of practical multivariable feedback systems. It starts from the following simple relationship. Assuming that the combined controller-and-plant forward path operator $Q(s)$ is non-singular, which is necessary if the m components of $y(s)$ are to respond independently to the m components of $r(s)$, we may write the inverse of the overall closed-loop transfer-function matrix as

$$R^{-1}(s) = Q^{-1}(s) + H(s) \qquad (11\text{-}21)$$

Inverse Nyquist Array (INA). - The set of m^2 loci in the complex plane corresponding to the entries of $Q^{-1}(j\omega)$ is called the inverse Nyquist array for the square matrix $Q(s)$.

If $H(s) = I_m$ then equation (11-21) gives a set of very simple relationships between open and closed-loop transfer function matrices. It is convenient, for the purpose of referring to the elements of $R^{-1}(s)$ and $Q^{-1}(s)$ to write

$$R^{-1}(s) = \hat{R}(s) \text{ and } Q^{-1}(s) = \hat{Q}(s) \qquad (11\text{-}22)$$

(because, in general, $q_{ij}^{-1}(s) \neq \hat{q}_{ij}(s)$ where q_{ij} and $\hat{q}_{ij}$ denote individual elements of the matrices Q and $\hat{Q}$; that is to say the elements of the inverse matrix are not in general inverses of the elements of the original matrix). Re-writing (11-21) in terms of the elements of the inverse matrices concerned gives

$$\hat{r}_{ii}(s) = \hat{q}_{ii}(s) + 1 \qquad (11\text{-}23)$$

$$\hat{r}_{ij}(s) = \hat{q}_{ij}(s) \qquad (11\text{-}24)$$

for $i,j = 1,2, \ldots, m$

Furthermore, if the jth feedback loop is opened (remember H is diagonal, but K is not), then we get

$$\hat{r}_{jj}(s) = \hat{q}_{jj}(s) \qquad (11\text{-}25)$$

Thus, in terms of frequency-response plots, the Inverse Nyquist Array allows the elements of $R^{-1}(j\omega)$ to be obtained in a simple way, whether some of the feedback loops are opened or not. The basic result on which the design technique is based is the following stability theorem due to Rosenbrock (Rosenbrock, 1969).

Rosenbrock's Stability Theorem. - A feedback system of the type shown in Figure 11-6 will be closed-loop stable if the system is opened-loop stable and if the controller matrix $K(s)$ is designed such that:

(i) a conventional inverse Nyquist mapping of each diagonal element

$\hat{q}_{ii}(j\omega)$ [i = 1,2, . . ., m] of the inverse Nyquist array for $Q(j\omega)$ encircles the critical point (-1,0) and the origin the same number of times in the same direction;

$$\text{(ii)} \quad |\hat{q}_{ii}(s)| > \sum_{\substack{j=1 \\ j\neq i}}^{m} |q_{ij}(s)| \quad \text{on D, the contour defined in Section 3.1} \qquad (11\text{-}26)$$

$$\text{(iii)} \quad |\hat{r}_{ii}(s)| > \sum_{\substack{j=1 \\ j\neq i}}^{m} |\hat{r}_{ij}(s)| \quad \text{on D} \qquad (11\text{-}27)$$

The condition imposed by (ii) and (iii) on $Q(s)$ and $R(s)$ respectively is called <u>diagonal dominance</u>. Making $Q(s)$ and $R(s)$ diagonal-dominated ensures that interactive effects are "small" in a precisely defined sense which enables stability of the closed-loop system to be determined from the diagonal elements of the open-loop transmission. This enables a designer to have the essential simplicity of a fully non-interacting approach from the feedback design point of view without requiring an excessively complicated controller or sacrificing too much design freedom in choosing the controller elements.

This stability criterion is only a sufficient one, since failure to satisfy conditions (ii) and (iii) does not infer closed-loop instability, regardless of (i). However, if (ii) and (iii) are satisfied, then (i) is both necessary and sufficient. This being the case, m single-loop controllers can be designed on the basis of a band in the complex plane swept out by the union of circles of radius and centre given by

$$\text{Radius} = \sum_{\substack{j=1 \\ j\neq i}}^{m} |\hat{q}_{ij}(j\omega)|$$

$$\text{Centre} = q_{ii}(j\omega)$$

The latest form of the method allows the magnitude of the circles to be reduced for compensation purposes, retaining the original size only for stability analysis (Rosenbrock, 1970). The method has also been extended to deal with open-loop unstable systems (McMorran, 1971).

The way in which the technique uses a band in the complex plane is important not only for stability determination but also for the following reason. The elements $r_{ii}(s)$ do not represent anything directly measurable on a physical plant, but the actual closed-loop inverse transferences denoted by $r_{ii}^{-1}(s)$ <u>are</u> physically measurable. Rosenbrock (Rosenbrock, 1969, 1970) has shown that these quantities $r_{ii}^{-1}(s)$ i = 1,2, . . ., m are trapped within the narrower bands defined above for all values of gain in each loop between zero and designed values, where zero gain in any one loop is equivalent to opening that loop. This has the most important consequence that the inverse Nyquist array technique, by virtue of the imposed diagonal dominance, is a high integrity design technique insofar as transducer failures are concerned, since these can be interpreted as opening loops. Also, when the method assumes $H(s) = I_m$, high integrity against error-monitoring channel failures is assured. However, in its present form, the method does not incorporate any form of check or protection against actuator failures (that is those occurring directly at the plant input).

When diagonal-dominance can be achieved by the use of an acceptable pre-controller, the inverse Nyquist array method is an extremely powerful technique. One of the main attractions of the method is that, after achievement of diagonal-dominance by design of a suitable pre-controller, the remainder of the design is completed on the basis of a set of individual single loops where the information concerning a particular loop is contained in a previously defined band regardless of what is happening in the other loops.

11.6 Characteristic Locus Technique

A design technique is being developed by MacFarlane and Belletrutti which is based on the stability, integrity and performance properties of the characteristic loci. (MacFarlane, 1971) (MacFarlane and Belletrutti, 1971). As the underlying ideas are explained in another paper in these conference proceedings, they will only be briefly discussed here. This method requires the use of a computer-aided design facility with a graphic display on which appropriate characteristic loci are computed and displayed. Instead of the bands which define an approximate location for stability purposes, as in the Inverse Nyquist Array method, an exact indication of stability is obtained from the appropriate set of characteristic loci. By using a number of return-difference matrices and their appropriate sets of characteristic loci, a complete integrity study may be incorporated into the design procedure. It appears from preliminary studies that the loci give a great deal of useful insight into the complicated process of choosing the entries in a multivariable controller matrix.

12. CONCLUDING REMARKS

The wealth of papers submitted to this conference shows that research in multivariable control systems is still in a very vigorous phase of development. A striking feature of the papers in the session on linear theory is the increasing use of modern algebraic techniques. The paper by Wonham and Morse (Group 1.1) on feedback invariants of linear multivariable systems is a notable example, which studies the invariant structure of a controllable matrix pair under a transformation group. This reflects the current abstract algebraic trend in studies of systems theory being conducted, for example, by Kalman (Kalman, 1965, 1966, 1967, 1968, 1969) and Rosenbrock (Rosenbrock, 1966, 1968, 1969, 1970). Such studies are closely related to algebraic investigations in the field of automata theory (Krohn and Rhodes, 1965), (Arbib, 1965), and it may well be that the most sophisticated linear algebraic system theories will have their greatest impact in the digital processing field, since in any model for an industrial dynamical system the "fine structure" of a high-order linear model rapidly becomes overwhelmed by non-linearities and uncertainties of structure and measurement in the practical system. Many papers in the sessions on linear system theory cover aspects of algebraic theory which are related to practical design considerations, such as the studies of decoupling by Cremer (Group 1.1); of sensitivity by Cruz and Sundarajan (Group 1.4); Ohno (Group 1.4) and Perkins (Group 1.5); of the construction of models by Gerth (Group 1.2) and Wolovich (Group 1.2); of controllability and observability by Muller and Weber (Group 1.1). Only a few papers are concerned with the design process as such. Chen (Group 1.3) describes a frequency-domain design technique for compensators for systems with single-input and multiple outputs. MacFarlane (Group 1.3) describes a frequency-domain design technique using characteristic loci for interacting systems, and Niederlinski (Group 1.3) describes a quantitative measure of interaction for multivariable plants. Grasselli and Nicola (Group 1.1) describe a modal synthesis technique for multivariable regulators.

Practising design engineers need to base their work on a few simple ideas of great intuitive appeal. The great success of the classical Bode-Nyquist methods largely stemmed from the establishment of such ideas for single-input single-output systems. There seems good reason to hope that the recent developments of multivariable frequency-response methods will make many recent theoretical advances accessible to practising engineers faced with the design of interactive multivariable systems.

The general features of linear multivariable feedback systems are now well established, and the main emphases in multivariable systems research will now probably shift decisively towards nonlinear theory and to the complex structures involved in adaptive and hierarchical control. Following the success in multivariable control work of algebra, that other basic pillar of modern mathematics, topology, will probably soon start to exercise a strong influence on research approaches in these fields.

13. REFERENCES AND BIBLIOGRAPHY

The text references give author and year only; in those cases where an author has several cited references given for the same year, the relevant one should be clear from the title quoted. The bibliography makes no claim to an exhaustive coverage or to historical completeness. Nevertheless, it is hoped that a significant proportion of recent research work in the field has been cited.

AMARA, R.C. 1959. "The linear least squares synthesis of multivariable control systems," Trans. AIEE, 78, II, pp. 115-119
AMARA, R.C. 1959. "Application of matrix methods to the linear least square synthesis of multivariable systems," J. Franklin Inst., 268, pp. 1-6.
ANDERSON, B.D.O. 1966. "The inverse problem of optimal control," Report No. SEL-66-038 (TR No. 6560-3) Stanford Electronics Laboratories, Stanford, Calif.
ANDERSON, B.D.O. 1966. "Sensitivity improvement using optimal design," Proc. IEE, 113, pp. 1084-1086.
ANDERSON, B.D.O. 1967. "An algebraic solution to the spectral factorization problem," IEEE Tr on A.C. Vol. AC-12, No. 4, pp. 410-414.
ANDERSON, B.D.O. 1967. "A system theory criterion for positive real matrices," SIAM J. Control, 5, pp. 171-182.
ANDERSON, B.D.O. and LUENBERGER, D.G. 1967. "Design of multivariable feedback systems," Proc. IEE, 114, (3), pp. 395-399.

ANDERSON, B.D.O. and MOORE, J.B. 1969. "Linear system optimisation with prescribed degree of stability," Proc. IEE, 116 (12), pp. 2083-2087.
ANDERSON, B.D.O. and NEWCOMB, R.W. 1968. Impedance synthesis via state-space techniques," Proc. IEE, 115 (7), pp. 928-936.
ARBIB, M.A. 1965. "A common framework for automata theory and control theory," SIAM Journal Control, Vol. 3, pp. 206-222.
ASSEO, S.J. 1968. "Application of optimal control to 'perfect' model following," Reprints of 1968 JACC, pp. 1056-1070.
ASSEO, S.J. 1968. "Phase-variable canonical transformation of multicontroller systems," IEEE Trans., AC-13 (1), pp. 129-131.
ATHANS, M. 1966. "The status of optimal control theory and applications for deterministic systems," Trans. IEEE, AC-11, pp. 580-596.
ATHANS, M. and FALB, P.L. 1966. "Optimal control," McGraw-Hill.

BARNETT, S. and STOREY, C. 1970. "Matrix methods in stability theory," Nelson, London.
BASILE, C. and MARRO, G. 1970. "A state space approach to non-interacting controls," Ricerche di Automatica, 1, pp. 68-77.
BASS, R.W. and GURA, I. 1965. "High-order system design via state-space considerations," JACC, pp. 311-318.
BELLMAN, R., GLICKSBERG, I. and GROSS, O. 1956. "On the bang-bang control problem," Quart. App. Math., 14, pp. 11-18.
BERKOVITZ, L.D. 1961. "Variational methods in problems of control and programming," Journal of Mathematical Analysis and its applications, 3, pp. 145-169.
BINGULAC, S.P. 1970. "An alternative approach to expanding PA + A'P = -Q," IEEE Trans., AC-15 (1), pp. 135-137.
BODE, H.W. 1945. "Network analysis and feedback amplifier design," Van Nostrand.
BOHN, E.V. 1960. "Stabilization of linear multivariable feedback control systems," Trans. IRE on Auto. Control, AC-5, pp. 321-327.
BOHN, E.V. 1962. "Design and synthesis methods for a class of multivariable feedback control systems based on single-variable methods," Trans. AIEE, 81, II, pp. 109-116.
BOKSENBOM, A.S. and HOOD, R. 1949. "General algebraic method applied to control analysis of complex engine types," Report NCA-TR-980, National Advisory Committee for Aeronautics, Washington, D.C.
BONGIORNO, J.J. Jr. and YOULA, D.C. 1968. "On observers in multivariable control systems," Int.J.Control, 8, pp. 221-243.
BONGIORNO, J.J. Jr. 1969. "Minimum sensitivity design of linear multivariable feedback control systems by matrix spectral factorization," IEEE Trans., AC-14 (6), pp. 665-673.
BRASCH, F.M. Jr. and PEARSON, J.B. 1970. "Pole placement using dynamic compensators," IEEE Trans., AC-15 (1), pp. 34-43.
BRISTOL, E.H. 1967. "A philosophy for single loop controllers in a multiloop world," Proceedings of the 8th National Symposium on Instrumentation in the Chemical and Process Industries, V.4, pp. 19-29.
BROCKETT, R.W. and MESAROVIC, M.D. 1962. "Synthesis of linear multivariable systems," Trans. AIEE, 81, II, pp. 216-221.
BROCKET, R.W. 1965. "Poles, zeros and feedback : State space interpretation," Trans. IEEE, AC-10, pp. 129-135.

BROCKETT, R.W. 1966. "The status of stability theory for deterministic systems," Trans. IEEE, AC-11, pp. 596-606.
BROCKETT, R.W. 1970. "Finite dimensional linear systems," New York, John Wiley & Sons.
BRUNI, C., ISIDORI, A. and RUBERTI, A. 1969. "A method of realization based on the moments of the impulse- response matrix," IEEE Trans., AC-14 (2), pp. 203-204.
BRYSON, A.E. and DENHAM, W.F. 1962. "A steepest-ascent method for solving optimum programming problems," Journal of Applied Mechanics, pp. 247-257.
BRYSON, A.E. and HO, Y-C. 1969. "Applied optimal control," Blaisdel, Waltham, Mass.
BRYSON, A.E. Jr. and LUENBEGER, D.G. 1970. "The synthesis of regulator logic using state-variable concepts," Proc. IEEE, 58, pp. 1803-1811.

CALAHAN, D.A. 1964. "Modern network synthesis," Hayden, New York.
CSAKI, F. and FISCHER, P. 1967. "On the spectrum factorization," Acta Technica Academiae Scientiarum Hungariae Tomus 58(1-2) pp. 145-168.
CHANG, S.S.L. 1961. "Synthesis of optimum control systems," McGraw-Hill, New York.
CHATTERJEE, H.K. 1960. "Multivariable process control," IFAC Congress, Moscow.
CHEN, C.F. and TSANG, N.F. 1967. "A stability criterion based on the return difference," IEEE Trans., E-10, pp. 180-182.
CHEN, C-T, and DESOER, C.A. 1967. "Controllability and observability of composite systems," IEEE Trans., AC-12 (4), pp. 402-409.
CHEN, C-T. 1968. "Representations of linear time-invariant composite systems," IEEE Trans., AC-13 (3), pp. 277-283.
CHEN, C-T. 1968. "Stability of linear multivariable feedback systems," Proc. IEEE, 56 (5), pp. 821-828.
CHEN, C-T, 1968. "Synthesis of multivariable feedback systems," Joint Automatic Control Conference, pp. 224-228.
CHEN, C-T. 1970. "Irreducibility of dynamial equation realizations of sets of differential equations," IEEE Trans. on AC - Vol AC 13, p. 131.
CHEN, R. and SHEN, D. 1968. "Computer-aided design of multivariable control systems with applications to VTOL aircraft stability augmentation," IFAC Symposium, Dusseldorf, General and Special Computer Procedures.
CLAERBOUT, J.F. 1966. "Spectral factorization of multivariable time series," Biometrika, Vol. 53, pp. 264-267.
CROSSLEY, T.R. and PORTER, B. 1969. "Eigenvalue and eigenvector sensitivities in linear systems theory," Int.J.Control, 10 (2), pp.163-170.
CROSSLEY, T.R. and PORTER, B. 1969. "Synthesis of aircraft control systems having real or complex eigenvalues," The Aeronautical J. of the Royal Aeronatuical Soc., 73 (698), pp. 138-142.
CRUZ, J.B. and PERKINS, W.R. 1964. "The role of sensitivity in the design of multivariable linear systems," Proc. NEC, Vol. 20, pp. 742-745.
CRUZ, J.B. and PERKINS, W.R. 1964. "A new approach to the sensitivity problem in multivariable feedback system design," Trans. IEE, AC-9, pp. 216-223.

CRUZ, J.B. and PERKINS, W.R. 1965. "The parameter variation problem in state feedback control systems," Trans. ASME, 87, D, pp. 120-124.
CRUZ, J.B. and PERKINS, W.R. 1965. "Sensitivity comparison of open-loop and closed-loop systems," Proc. Third Allerton Conf. on Circuit and System Theory, University of Illinois, Urbana, pp. 607-612.
CRUZ, J.B. and PERKINS, W.R. 1966. "Criteria for system sensitivity to parameter variations," Proc. Third Congress of IFAC, London, pp. 18C.1-18C.7.
CUMMING, S.D.G. 1969. "Design of observers of reduced dynamics," Electron. Lett., 5 (10), pp. 213-214.
CUMMING, S.D.G. 1969. Observer theory and control system design," Ph.D. thesis, Centre for Computing and Automation, Imperial College, Univ. of London.

DABKE, K.P. 1970. "Suboptimal linear regulators with incomplete state feedback," IEEE Trans., AC-15 (1), pp. 120-122.
DAVIS, M.C. 1963. "Factoring the spectral matrix," IEEE Trans. on A.C. Vol. AC-8, pp. 296-305.
DAVISON, E.J. 1967. "Control of a distillation column with pressure variation," Trans. Instn. Chem. Engrs., Vol. 45 No. 6, pp. T279-T250.
DAVISON, E.J. and WONHAM, W.M. 1968. "On pole assignment in multivariable linear systems," IEEE Trans., AC-13 (6), pp. 747-748.
DAVISON, E.J. and GOLDBERG, R.W. 1968. "Optimum variable measurement in multivariable control systems," IFAC Symposium, Dusseldorf. Applications - Power Plants.
DAVISON, E.J. 1969. "A non minimal phase index and its application to interactive multivariable control systems," 4th IFAC Congress, Session 61, pp. 3-21. Warsaw.
D'AZZO, JJ. and HOUPIS, C.H. 1966. "Feedback control system analysis and synthesis," McGraw-Hill.
DELLON, F. and SARACHIK, P.E. 1968. "Optimal control of unstable linear plants with inaccessible states," Trans. IEE, AC-13, pp. 491-495.
DESOER, C.A. "The bang-bang servo problem treated by variational techniques," Information and Control, 2, pp. 333-348.
DESOER, C.A. 1961. "Pontagin's maximum principle and the principle of optimality," Journal of Franklin Institute, 271, pp. 361-367.
DESOER, C.A. and CHEN, C-T. 1967. "Controllability and observability of feedback systems," IEEE Trans., AC-12 (4), pp. 474-475.
DIXON, B.A. 1971. "A critical assessment of linear multivariable feedback control system design techniques," Ph.D. thesis, Control Systems Centre, UMIST.

ELLIS, J.K. and WHITE, G.W.T. 1965. "An introduction to modal analysis and control," CONTROL, April, May, June, 1965, pp. 193-197, 262-266, 317-321.
ERZBERGER, H. 1968. "Analysis and design of model following control systems by state space techniques," JACC 1968, p. 572-581
EVANS, W.R. 1967. "Graphical analysis control systems," Trans. AIEE, 67, pt.I, pp. 547-551.

FALB, P.L. and WOLOVICH, W.A. 1967. "Decoupling in the design and synthesis of multivariable control systems," IEEE Trans., AC-12 (6) pp. 651-659.
FERGUSON, J.D. and REKASIUS, Z.V. 1969. "Optimal linear control systems with incomplete state feedback," IEEE Trans., AC-14 (2), pp. 135-140.
FRANCIS, N.D. 1967. "Sensitivity analysis in multivariable system design," Electron. Lett., 3 (9), pp. 404-405.
FREEMAN, H. 1957. "A synthesis method for multipole control systems," Trans. AIEE, 76, II, pp. 28-31.
FREEMAN, H. 1958. "Consideration of synthesis of multipole control systems," AIEE Trans. on Appl. & Industry, pp. 1-5.
FREEMAN, H. 1958. "Stability and physical realizability considerations in the synthesis of multiple control systems," Trans. AIEE, Part II, pp. 1-5.
FREEMAN, H. 1965. "Discrete-time systems," Wiley.
FULLER, A.T. 1962. "Bibliography of optimum nonlinear control of determinate and stochastic-definite systems," Journal of Electronics and Control, 13. pp. 589-611.

GANTMACHER, F.R. 1959. "Theory of matrices," Vol. 1 and 2, Chelsea, New York.
GARDNER, M.R. ROSS ASHBY, W. 1970. "Connectance of large dynamic/cybernetic/systems: critical values for stability," Nature 228, p. 784.
GELB, A. and VANDER VELDE, W.E. 1968. "Multiple-input describing functions and non-linear system design," McGraw-Hill.
GILBERT, E.G. 1963. "Controllability and observability in multivariable control systems," J. SIAM on Control. Ser.A, 1 (2), pp. 128-151
GILBERT, E.G. 1969. "The decoupling of multivariable systems by state feedback," J.SIAM on Control, 7 (1), pp. 50-63.
GILBERT, E.G. and PIVNICHNY, J.R. 1969. "A computer program for the synthesis of decoupled multivariable feedback systems," IEEE Trans., AC-14 (6), pp. 652-659.
GILLE, J-C, PELEGRIN, M.J. and DECAULNE, P. 1959. "Feedback control systems," McGraw-Hill, New York.
GOLOMB, M. and USDIN, E. 1952. "A theory of multidimensional servo systems," J. Franklin Inst., 253, pp. 29-58.
GORDON-CLARK, M.R. 1964. "A novel approach to the control of dynamically unfavourable processes," IEEE Trans. on Automatic Control, pp. 411-419.
GOULD, L.A., MURPHY, A.T. and BERKMAN, E.F. 1970. "On the Simon-Mitter pole allocation algorithm - explicit gains for repeated eigenvalues," IEEE Trans., AC-15 (2), pp. 259-260.
GUEGUEN,C.J. and TOUMIRE, E., PANDA, S.P. and CHEN, C-T. 1969. "Comments on 'Irreducible Jordan form realization of a rational matrix," IEEE Trans. AC-14 (6), pp. 783-784.
GUEGUEN, C.J. 1970. "Proprietes fondamentales de la representation des systemes multidimensionnels par operateurs differentiels," C.R. Acad. Sc. Paris, t 271, pp. 858-860.
GUEGUEN, C.J. TOUMIRE, E. 1968. "Representations minimales en variables d'etat," Automatisme. Vol. 13, No. 5, pp. 197-203.
GUPTA, S.C. 1966. "Transform and state variable methods in linear systems," Wiley.

HALKIN, H. 1963. "The principle of optimal evolution"in Nonlinear differential equations and nonlinear mechanics, Academic Press.
HARRIOTT, P. 1964. "Process Control," McGraw-Hill Book Co., New York, pp. 100-102.
HARRISON, H.L. and BOLLINGER, J.G. 1963. "Introduction to automatic controls," International Textbook Company, Scranton, Pennsylvania.
HERSTEIN, I.N. 1964. "Topics in algebra," Blaisdell.
HEYMANN, M. and WONHAM, W.M. 1968. "On pole assignment in multi-input controllable linear systems," IEEE Trans., AC-13 (6), pp. 748-749.
HEYMANN, M. and THORPE, J.A. 1970. "Transfer equivalence of linear dynamical systems," SIAM J. Control, 8, pp. 19-40.
HO, B.L. and KALMAN, R.E. 1965. "Effective construction of linear state variable models from input/output data," Proc. 3rd Annual Allerton Conf. on Circuit and System Theory, pp. 449-459.
HODGE, S.S. 1970. "Regulation of the reheat system of a jet engine," Ph.D. thesis, Control Systems Centre, UMIST.
HOROWITZ, I.M. 1960. "Synthesis of linear multivariable feedback control systems," Trans. IRE, AC-5, pp. 94-105.
HOROWITZ, I.M. 1963. "Synthesis of feedback systems," Academic Press, New York.
HSU, C.H. and CHEN, C.T. 1968. "A proof of the stability of multivariable feedback systems," Proc. Inst. Elect. Electron. Engrs., 56, pp. 2061-2062.

JAMES, H.M., NICHOLS, N.B. and PHILLIPS, R.S. 1947. "Theory of servomechanisms," McGraw-Hill.
JAMESON, A. 1968. "Solution of the equation AX + XB = C by the inversion of an MxM or NxN matrix," SIAM J. Appl. Math., 16, pp. 1020-1023.
JANSSEN, J.M.L. 1965. "Control systems behaviour expressed as a deviation ratio," in R. Oldenbourger "Frequency Response," The Macmillan Co., New York, pp. 131-140.
JOHNSON, C.D. and GIBSON, J.E. 1963. "Singular solutions in problems of optimal control," Trans. IEEE on Automatic Control, AC-8, pp. 4-15.

KALMAN, R.E. and BERTRAM, J.E. 1958. "General synthesis procedure for computer control of single and mutli-loop linear systems," Trans. AIEE, Vol. 77, III, pp. 602-609.
KALMAN, R.E. and KOEPCKE, R.W. 1958. "Optimal synthesis of linear sampling control systems using generalized performance indexes," Trans. ASME, 80, pp. 1820.
KALMAN, R.E. and BERTRAM, J.E. 1960. "Control system analysis and design via the 'second' method of Lyapunov. Part I Continuous-Time Systems," Trans. ASME, 82, D, pp. 371-393.
KALMAN, R.E. 1960. "On the general theory of control systems," Prof. IFAC, Moscow, 1, London, Butterworth.
KALMAN, R.E. 1960. "Contributions to the theory of optimal control," Proceedings of Mexico City Conference on Ordinary Differential Equations, 1959, also in: Bol. Soc. Mat. Mex., Second series, Vol, 5, 1960, pp. 102-119.
KALMAN, R.E. and BUCY, R.S. 1961. "New results in linear filtering and prediction theory," Trans. ASME, Ser. D, J. Basic Engrg., pp. 95-108.

KALMAN, R.E. 1962. "Canonical structure of linear dynamical systems," Proc. Nat. Acad. Sci., Vol. 48, No. 4, pp. 596-600.
KALMAN, R.E. 1963. "Liapunov functions for the problem of Lur'e in automatic control," Proc. NAS (USA), 49, pp. 201-205.
KALMAN, R.E. 1963. "Mathematical description of linear systems," J. SIAM on Control, Ser. A., 1 (2), pp. 152-192.
KALMAN, R.E., HO, Y.C. and NARENDA, K.S. 1963. "Controllability of linear dynamical systems," Contr. Diff. Eqns. 1 (2), pp. 189-213.
KALMAN, R.E. 1963. "The theory of optimal control and the calculus of variations," Mathematical Optimization Techniques, Univ. of California Press, Berkeley.
KALMAN, R.E. 1964. "When is a linear system optimal?" ASME Trans., J. of Basic Engng.(D), 86, pp. 51-60.
KALMAN, R.E. 1964. "Towards a theory of difficulty of computation in optimal control," Proceedings IBM Symposium on Control Theory and Applications, Thomas J. Watson Research Center, Yorktown Heights, New York.
KALMAN, R.E. 1965. "Algebraic structure of linear dynamical systems 1. The module of Σ," Proc. Nat. Acad. Sc., Vol. 54, No. 6, pp. 1503-1508.
KALMAN, R.E. 1965. "Algebraic theory of linear systems," Proc. Third Allerton Conference on Circuit and System Theory, pp. 563-577.
KALMAN, R.E. 1965. "Irreducible realizations and the degree of a rational matrix," J. SIAM, 13 (2), pp. 520-544.
KALMAN, R.E. 1966. "On the structural properties of linear, constant multivariable systems," Proc. Third IFAC Congress, London.
KALMAN, R.E. 1966. "New developments in systems theory relevant to biology," Third Systems Symposium, "Systems Approach in Biology," Case Institute of Technology, Cleveland, Ohio.
KALMAN, R.E. "Algebraic aspects of the theory of dynamical systems," in "Differential Equations and Dynamical Systems," edited by J.K. Hale and J.P. La Salle, Academic Press, New York.
KALMAN, R.E. 1967. "On the mathematics of model building," Proceedings of School on Neural Networks, Ravello.
KALMAN, R.E. 1968. "Lectures on controllability and observability," C.I.M.E. Seminar on Controllability and Observability, Bologna.
KALMAN, R.E. 1969. "Algebraic characterization of polynomials whose zeros lie in certain algebraic domains," Proc. Nat. Acad. Sc., Vol. 64, No. 3, pp. 818-823.
KALMAN, R.E. 1969. "Introduction to the algebraic theory of linear dynamical systems," Lecture Notes in Operations Research and Mathematical Economics, Vol. 11, Springer-Verlag, Berlin.
KALMAN, R.E., FALB, P.L. and ARBIB, M.A. 1969. "Topics in mathematical system theory," McGraw-Hill.
KATKOVNIK, V. Ya. and POLUEKTOV, R.A. 1965. "The synthesis problem for optimal multidimensional automatic control systems," Auto. and Rem. Cont., 26(1), pp. 17-26.
KATKOVNIK, V. Ya. 1966. "To the theory of synthesis of the multidimensional linear control systems," Proc. 3rd IFAC, VI, paper 1D.
KAVANAGH, R.J. 1957. "Noninteraction in linear multivariable systems," Trans. AIEE, 76, II, pp. 95-100.
KAVANAGH, R.J. 1957. "The application of matrix methods to multivariable control systems," J. Franklin Inst., 262, pp. 349-367.
KAVANAGH, R.J. 1958. "Multivariable control system synthesis," Trans. AIEE, 77, II, pp. 425-429.

KAVANAGH, R.J. 1961. "A note on optimum linear multivariable filters," Proc. IEE, 108C, pp. 412-417.
KAVANAGH, R.J. 1968. "Spectral-matrix factorization," Electronics Letters, Vol. 4, No. 24, pp. 527-528.
KESAVAN, H.K., SARMA, I.G. and PRASAD, U.R. 1969. "Sensitivity-state models for linear systems," Int.J. Control, 9 (3), pp. 291-310.
KINNEN, E. and LIU, D.S. 1962. "Linear multivariable control system design with root loci," Trans. AIEE, 81, II, pp. 41-44.
KLEINMAN, D.L. 1968. "On an iterative technique for Riccati equation computations," IEEE Trans., AC-13 (1), pp. 114-115.
KOIVO, A.J. 1970. "Performance sensitivity of dynamical systems," Proc. IEE, 117 (4), pp. 825-830.
KOIVUNIEMI, A.J. 1967. "A unified approach to sensitivity of discrete-time systems," Proc. 1st Ann. Princeton Conf. Eng. Sci. Syst., Princeton University, Princeton, N.J., pp. 12-15.
KOKOTOVIC, P. and RUTMAN, R.S. 1965. "Sensitivity of automatic control systems (survey)," Auto and Rem. Cont., 26(1), pp. 727-749.
KREINDLER, E. 1968. "Closed-loop sensitivity reduction of linear optimal control systems," Trans. IEEE, AC-13, pp. 254-262.
KREINDLER, E. 1968. "On the definition and application of the sensitivity function," Journal Franklin Institute, Vol. 285, No. 1, pp. 26-36.
KROHN, K.B. and RHODES, J.L. 1965. "Algebraic theory of machines. 1. The main decomposition theorem," Trans. Amer. Math. Soc., Vol. 116, pp. 450-464.
KUO, B.C. 1963. "Analysis and synthesis of sampled-data control systems," Prentice-Hall, Englewood Cliffs, N.J.

LASALLE, J.P. 1959. "The time-optimal control problem," Contributions to the Theory of Nonlinear Oscillations, Vol. 5, pp. 1-24.
LASALLE, J.P. and LEFSCHETZ, S. 1961. "Stability by Liapunov's direct method with applications," Academic, New York.
LEAKE, R.J. 1965. "Return difference Bode diagram for optimal system design," IEEE Trans. AC-10, pp. 342-344.
LEE, R.C.K. 1964. "Optimal estimation, identification and control," (M.I.T. Press).
LEITMANN, G. 1966. "An introduction to optimal control," McGraw-Hill.
LEVINE, W.S. and ATHANS, M. 1966. "On the optimal error regulation of a string of moving vehicles," Trans. IEEE AC-11, pp. 355-361.
LINDGREN, A.G. 1966. "A note on the stability and design of interacting control systems," IEEE Trans., AC-11 (2), pp. 314-315.
LIU, C.K. and BERGMAN, N.J. 1970. "On necessary conditions for decoupling multivariable systems," IEEE Trans., AC-15 (1), pp. 131-133.
LOO, S.G. 1967. "Spectral factorization by means of augmented factors," Electronics Letters, Vol. 3, No. 6, pp. 238-239.
LUENBERGER, D.G. 1964. "Observing the state of a linear system," IEEE Trans., MIL-8, pp. 74-80.
LUENBERGER, D.G. 1966. "Observers for multivariable systems," IEEE Trans., AC-11 (2), pp. 190-197.
LUENBERGER, D.G. 1967. "Canonical forms for linear multivariable systems," IEEE Trans., AC-12 (3), pp. 290-293.

MACFARLANE, A.G.J. 1963. "An eigenvector solution of the optimal linear regulator problem," Journal of Electronics and Control, Vol. 14, No. 6, pp. 643-654.
MACFARLANE, A.G.J. 1964. "Engineering systems analysis," Harrap.
MACFARLANE, A.G.J. 1968. "System matrices," Proc. IEE, 115, pp. 749-754.
MACFARLANE, A.G.J. 1968. "Use of moving frames in the representation of nonlinear and nonplanar state-space flows," Proc. IEE, Vol. 115, No. 8, pp. 1200-1206.
MACFARLANE, A.G.J. 1968. "Representation od state-space flows by circles in the complex plane," Proc. IEE, Vol. 115, No. 8, pp. 1195-1199.
MACFARLANE, A.G.J. and MUNRO, N. 1968. "Mappings of the state space into the complex plane and their use in multivariable systems analysis," Int. Journ. Control, Vol. 7, No. 6, pp. 501-555.
MACFARLANE, A.G.J. 1969. "Dual-system methods in dynamical analysis, Part 2 - Optimal regulators and optimal servomechanisms," Proc. IEE, 116 (8) pp. 1458-1462.
MACFARLANE, A.G.J. and MUNRO, N. 1969. "Use of generalized Mohr circles for multivariable regulator design," Proceedings of the 4th international IFAC congress, Warsaw, Paper 61.3.
MACFARLANE, A.G.J. 1969. "Use of power and energy concepts in the analysis of multivariable feedback controllers," Proc. IEE, 116, (8) pp. 1449-1452.
MACFARLANE, A.G.J. 1969. "Dual system methods in dynamical analysis. Pt. 1 - Varational principles and their application to nonlinear network theory," Proc. IEE, 116, (8) pp. 1452-1457.
MACFARLANE, A.G.J. 1970. "Dynamical System Models," Harrap.
MACFARLANE, A.G.J. 1970. "Commutative controller : a new technique for the design of multivariable control systems," Electron. Lett., 6 (5), pp. 121-123.
MACFARLANE, A.G.J. 1970. "Two necessary conditions in the frequency domain for the optimality of a multiple-input linear control system," Proc. IEE, 117, (2), pp. 464-466.
MACFARLANE, A.G.J. 1970. "The return-difference and return-ratio matrices and their use in the analysis and design of multivariable feedback control systems," Proc. IEE, 117, pp. 2037-2049.
MACFARLANE, A.G.J. 1970. "The use of characteristic functions and characteristic values in feedback systems analysis," Int. J. Math. Educ. Sci. Technol, Vol. 1, pp. 359-366.
MACFARLANE, A.G.J. and ROSENBROCK, H.H. 1970. "New vector-space structure for dynamical systems," Electron. Lett., 6 (6), pp. 162-163.
MACFARLANE, A.G.J. 1971. "Return-difference matrix properties for optimal stationary Kalman-Bucy filter," Proc. IEE, Vol. 118, pp. 373-376.
MACFARLANE, A.G.J. 1971. "Necessary conditions for a multivariable system to have high integrity," CSC Report No. 138.
MACFARLANE, A.G.J. 1971. "Use of characteristic transfer functions in the design of multivariable control systems," Proc. 2nd IFAC Conference on Multivariable Systems Theory, Dusseldorf.
MAN, F.T. 1970. "Suboptimal control of linear time-invariant systems with incomplete feedback," IEEE Trans., AC-15 (1), pp. 112-114.
MARKUS, L. and LEE, E.B. 1967. "Foundations of optimal control theory," Wiley.

MARSHALL, S.A. and NICHOLSON, H. 1970. "Optimal control of linear multivariable systems with quadratic performance criteria," Proc. IEE, Vol. 117, No. 8, pp. 1705-1713.
MAYNE, D.Q. 1968. "Computational procedure for the minimal realization of transfer-function matrices," Proc. IEE, 115 (9), pp. 1363-1368.
MAYNE, D.Q. and MURDOCH, P. 1970. "Modal control of linear time invariant systems," Int. J. Control, 11 (2), pp. 223-227.
MCBRIDE, L.E. and NARENDRA, S. 1963. "An expanded matrix representation for multivariable systems," IEEE Trans. ou A.C. Vo. AC 8.
MCCLAMROCH, N.H. 1969. "Evaluation of suboptimality and sensitivity in control and filtering processes," Trans. IEEE, AC-14, pp. 282-285.
MCMILLAN, B. 1952. "Introduction to formal realizability theory," Bell Syst. Tech. J., 31, pp. 217-279,541-600.
MCMORRAN, P.D. 1970. "Extension of the inverse Nyquist array," Electronics Letters, Vol. 6, No. 25, pp. 800-801.
MCMORRAN, P.D. 1970. "Design of gas turbine controller using inverse Nyquist method," Proc. IEE, 117, pp. 2050-2056.
MCMORRAN, P.D. 1970. "Parameter sensitivity and the inverse Nyquist method," Control Systems Centre Report No. 119, U.M.I.S.T.
MENAHEM, M. 1966. "Notes on the functional representation and the analysis of linear multidimensional processes," Proc. 3rd IFAC, Paper 1E.
MERRIAM, C.W. 1964. "Optimization theory and the design of feedback control systems," McGraw-Hill.
MESAROVIC, M.D. 1960. "The control of multivariable systems," MIT, Cambridge, Mass.
MITCHELL, D.S. and WEBB, C.R. 1960. "A study of interaction in multi-loop control system," IFAC Congress, Moscow.
MOLINARI, B.P. 1969. "Algebraic solution of matrix linear equations in control theory," Proc. IEE, 116 (10), pp. 1748-1754.
MORGAN, B.S. 1964. "The synthesis of linear multivariable systems by state variable feedback," Trans. IEEE on Auto. Control, Vol.AC-9, pp. 405-411.
MORGAN, B.S. Jr. 1966. "Sensitivity analysis and synthesis of multivariable systems," IEEE Trans., AC-11 (3), pp. 506-512.
MORSE, A.S. and WONHAM, W.M. 1970. "Decoupling and pole assignment by dynamic compensation," SIAM J. Control 8, pp. 317-338.
MUELLER, G.S. 1967. "An appraisal of the regulation of a turbojet," M.Sc. Dissertation, Control Systems Centre, UMIST.
MUFT, I.H. 1969. "A note on the decoupling of multivariable systems," IEEE Trans., AC-14 (4), pp. 415-416.
MUNRO, N. and MCMORRAN, P.D. 1970. "Signal-flow-graph reduction Mason's rule and the system matrix," Electron. Lett., 6, pp. 752-754.

NEWCOMB, R.W. 1966. "Linear multiport synthesis," McGraw-Hill, New York.
NEWCOMB, R.W. and ANDERSON, B.D.O. 1967. "A distributional approach to time-varying sensitivity," SIAM J. Appl. Math., 15, pp. 1001-1010.
NEWMANN, M.M. 1969. "Design algorithms for minimal-order Luenberger observers," Electron. Lett., 5 (17), pp. 390-392.
NEWMANN, M.M. 1969. "Optimal and suboptimal control using an observer when some of the state variables are not measurable," Int.J. Control, 9 (3), pp. 281-290.

NEWTON, G.C. Jr., GOULD, L.A. and KAISER, J. 1957. "Analytical design of linear feedback controls," Wiley, New York.
NICHOLSON, H. 1964. "Dynamic optimization of a boiler," Proc. IEE, 111, pp. 1479-1499.
NICHOLSON, H. 1967. "Eigenvalue and state-transition sensitivity of linear systems," Proc. IEE, 114 (12), pp. 1991-1995.
NIEDERLINSKI, A. 1970. "Analysis and design of 2-variable interacting control systems using inverse polar plots," Proc. IEE, 117, pp. 2056-2058.
NYQUIST, H. 1932. "Regeneration theory," Bell Syst. Tech. J., 11, pp. 126-147.

O'DONNELL, J.J. 1966. "Asymptotic solution of the matrix Riccati equation of optimal control," 4th Ann. Allerton Conf. on Circuit & System Theory, Univ. of Illinois, pp. 577-586.
OGATA, K. 1967. "State space analysis of control systems," Prentice-Hall, Englewood Cliffs, N.J.
OLDENBURGER, R. 1966. "Optimal and self-optimizing control," M.I.T. Press.
OSTROWSKI, A.M. 1952. "Note on bounds for determinants with dominant principal diagonal," Proc. Am. Math. Soc., Vol. 3, pp. 26-30.

PAIEWONSKY, B. 1965. "Optimal control: A review of theory and practice," AIAA Journal, 3, pp. 1985-2006.
PANDA, S.P. and CHEN, C-T. 1969. "Irreducible Jordan-form realization of a rational matrix," IEEE Trans., AC-14 (1), pp. 66-69.
PEARSON, J.B. and DING, C.Y. 1969. "Compensator design for multivariable linear systems," IEEE Trans., AC-14 (2), pp. 130-134.
PEARSON, J.B. 1969. "Compensator design for dynamic optimization," Int. J. Control, 9, pp. 473-482.
PERKINS, W.R. and CRUZ, J.B. 1965. "The parameter variation problem in state feedback control systems," ASME Trans. J. Basic Engng, Vo. 87, pp. 120-124.
PERKINS, W.R. and CRUZ, J.B. 1966. "Sensitivity operators for linear time-varying systems," in Sensitivity Methods in Control Theory, edited by L. Radanovic, Pergamon Press, New York, pp. 67-77.
PERKINS, W.R., CRUZ, J.B. Jr. and GONZALES, R.L. 1968. "Design of minimum sensitivity systems," Trans. IEEE, AC-13, pp. 159-167.
PERKINS, W.R. and D.F. WILKIE. 1970. "Controllability and observability of sensitivity models," Proc. 4th Ann. Princeton Conf. on Information Sciences and Systems, pp. 180-183.
PETROV, I.P. 1968. "Variational methods in optimum control theory," Academic Press.
POLAK, E. 1966. "An algorithm for reducing a linear time-invariant differential system to state-form, IEEE Trans. on A.C. Vol. AC-11, pp. 577-579.
PONTRYAGIN, L.S., BOLTYANSKY, V.G., GAMKRELIDZE, R.V. and MISCHENKO, Ye. F. 1963. "The mathematical theory of optimal processes," Interscience.
POPOV, V.M. 1964. "Hyperstability and optimality of automatic systems with several control functions," Revue Roumaine de Sciences Techniques - Electrotechnics and Energetics, 9 (4), Bucharest.

PORTER, B. and MICKLETHWAITE, D.A. 1967. "Design of multi-loop modal control systems," Trans. Soc. Inst. Tech., pp. 143-152.
PORTER, B. and CARTER, J.D. 1968. "Design of multi-loop modal control systems for plants having complex eigenvalues," Measurement and Control, Vol. 1, pp. T61-T68.
PORTER, B. and WOODHEAD, M.A. 1968. "Performance of optimal control when some of the state variables are not measurable," Int. J. Control, 8 (2), pp. 191-195.
PORTER, B. 1969. "Eigenvalue sensitivity of modal control systems to loop gain variations," Int. J. Control, 10 (2), pp. 159-162.
PORTER, W.A. 1965. "Sensitivity problems in linear systems," Trans. IEEE, AC-10, pp. 301-307.
PORTER, W.A. 1967. "On the reduction of sensitivity in multivariate systems," Int. J. Control 5, pp. 1-9.
PORTER, W.A. 1969. "An algorithm for inverting linear dynamic systems," Trans. IEEE, AC-14, pp. 702-704.
PORTER, W.A. 1969. "Decoupling of and inverses for time-varying linear systems," Trans. IEE, AC-14, pp. 378-380.
POTTER, J.E. 1966. "Matrix quadratic solution," SIAM Journal of Applied Mathematics, Vol. 14, No. 3, pp. 496-501.

RAE, W.G. 1964. "Synthesis of noninteracting control systems," Control, pp. 245-247.
REDDY, D.C. 1968. "Eigenfunctions and the solution of sensitivity equations," Electron. Lett., 4 (13), pp. 262-263.
RETALLACK, D.G. and MACFARLANE, A.G.J. 1969. "Analytical design procedures for a class of linear time-invariant multivariable regulators," Proc. IEE, 116 (6), pp. 1094-1100.
RETALLACK, D.G. 1970.. "Extended root-locus technique for the design of linear multivariable feedback systems," Proc. IEE, 117 (3), pp. 618-622.
RETALLACK, D.G. and MACFARLANE, A.G.J. 1970. "Pole-shifting techniques for multivariable feedback systems," Proc. IEE, 117 (5), pp. 1037-1038.
RETALLACK, D.G. 1970. "Transfer-function matrix approach to observer design," Proc. IEE, 117 (6), pp. 1153-1155.
RETALLACK, D.G. 1970. "Vector compensation and feedback theory," Ph.D. Thesis, Control Systems Centre, UMIST.
ROHRER, R.A. and SOBRAL, M. Jr. 1965. "Sensitivity considerations in optimal system design," Trans. IEEE, AC-10, pp. 43-48.
ROSENBROCK, H.H. 1962. "Control of a distillation column," Trans. Instn. Chem. Engrs. 40, Nr. 1, pp. 35-53.
ROSENBROCK, H.H. 1962. "Distinctive problems of process control," Chem.Eng. Progress, Vol. 58, No. 9, pp. 43-50.
ROSENBROCK, H.H. 1966. "On the design of linear multivariable control systems," Proc. 3rd Congress IFAC, London, VI, Paper 1A.
ROSENBROCK, H.H. 1967. "Least order of system matrices," Electronics Letters, Vol. 3, No. 2, pp. 58-59.
ROSENBROCK, H.H. 1967. "On linear system theory," Proc. IEE, 114 (11), pp. 1353-1359.
ROSENBROCK, H.H. 1968. "Computation of minimal representations of a rational transfer function matrix," Proc. IEE, 115, pp. 325-327.
ROSENBROCK, H.H. 1968. "Relatively prime polynomial matrices," Electronics Letters, No. 4, pp. 224-228.
ROSENBROCK, H.H. 1969. "Properties of linear constant systems," UMIST Control Systems Centre Report 63.

ROSENBROCK, H.H. 1969. "Design of multivariable control systems using the inverse Nyquist array," Proc. IEE, 116 (11) pp. 1929-1936.
ROSENBROCK, H.H. and ROWE, A. 1969. "Allocation of poles and zeros," Manchester Control Systems Centre Report No. 64, to be published in Proc. IEE.
ROSENBROCK, H.H. 1970. "Stability margins of closed-loop systems," Control Systems Centre Report No. 109, UMIST.
ROSENBROCK, H.H. and STOREY, C. 1970. "Mathematics of dynamical systems," Nelson.
ROSENBROCK, H.H. 1970. "State-space and multivariable theory, Nelson.
ROSENBROCK, H.H. 1971. "The stability of multivariable systems," Control Systems Centre Report No. 143, UMIST.
ROSENBROCK, H.H. and MCMORRAN, P.D. 1970. "State-space analysis of a cascaded controller," Proc. IEE, 117 (5), pp. 1026-1030.
ROSENBROCK, H.H. and MCMORRAN, P.D. 1971. "Good, bad or optimal?" Control Systems Centre Report No. 135, UMIST.
ROSENBROCK, H.H. 1971. "Progress in the design of multivariable control systems," Trans. Inst. Measurement and Control, Vol. 4, pp. 9-11.

SAGE, A.P. 1968. "Optimum systems control," Prentice-Hall.
SCHULTZ, D.G. and MELSA, J.L. 1967. "State functions and linear control systems," McGraw-Hill, New York.
SAIN, M.K. and MASSEY, J.L. 1969. "Inveribility of linear time-invariant dynamical systems," Trans. IEEE, AC-14, pp. 141-149.
SILVERMAN, L.M. and MEADOWS, H.E. 1967. "Controllability and observability in time-variable linear systems," J. SIAM on Control, 5, pp. 64-73.
SILVERMAN, L.M. and ANDERSON, B.D.O. 1968. "Controllability, observability and stability of linear systems," J. SIAM on Control, 6 (1), pp. 121-130.
SILVERMAN, L.M. 1969. "Inversion of multivariable linear systems," Trans. IEEE, AC-14, pp. 270-276.
SILVERMAN, L.M. 1970. "Decoupling with state feedback and precompensation," Trans. IEEE, AC-15, pp. 487-489.
SIMES, J.G. and ETZWEILER, G.A. 1967. "The design of multivariable linear systems by minimization of parameter induced error," Proc. JACC, pp. 575-580.
SIMON, J.D. 1967. "Theory and application of modal control," Systems Research Centre Report, SRC 104-A-67-46, Case Institute of Technology, U.S.A.
SIMON, J.D. and MITTER, S.K. 1968. "A theory of modal control," Information and Control, 13, pp. 316-353.
SIMON, J.D. and MITTER, S.K. 1969. "Synthesis of transfer function matrices with invariant zeros," IEEE Trans., AC-14 (4), pp. 420-421.
SIVAN, R. and SHALVI, R. 1968. "On the existence of optimal solutions for the linear system quadratic cost problem," Trans. IEEE, Vol. AC-13, pp. 188-191.
SOBRAL, M. Jr. 1968. "Sensitivity in optimal control systems," Proc. IEEE, 56, pp. 1644-1652.
STARKERMAN, R. and VOGL-FERNHEIM, E. 1968. "Investigations on the stability of control loop chains," IFAC Symposium on Multivariable Control Systems, Dusseldorf.

STONER, J.W., TAYLOR, F.J. and BASS, R.W. 1966. "Lateral stability augmentation of supersonic aircraft using multichannel optimal feedback control," ASS, Southeastern Symp. on Missiles and Aerosp. Veh. Sci., Huntsville, Proc. Vol. 2.

TASNY-TSCHIASSNY, L. 1953. "The return-difference matrix in linear networks," Proc. IEE, 100, Pt. IV, pp. 39-46.

THIEDE, E.C. and STUBBERUD, A.R. 1965. "Control system synthesis using linear transformations," IEEE Trans. AC-10, pp. 172-177.

TOU, J.T. 1959. "Digital and sampled-data control systems," McGraw-Hill, New York.

TRUXAL, J.G. 1955. "Automatic Feedback Control System Synthesis," McGraw-Hill, New York.

TSEIN, H.S. 1954. "Engineering Cybernetics," McGraw-Hill.

TYLER, J.S. 1964. "The characteristics of model-following systems as synthesised by optimal control," IEEE Trans. on A.C., Vol. AC-9.

TYLER, J.S. Jr. and TUTEUR, F.B. 1966. "The use of a quadratic performance index to design multivariable control systems," IEEE Trans., AC-11 (1), pp. 84-92.

VAUGHAN, D.R. 1968. "A negative exponential solution for the linear optimal regulator problem," JACC preprints, pp. 717-725, Univ. of Michigan.

VAUGHAN, D.R. 1969. "A negative exponential solution for the matrix Riccati equation," Trans. IEEE Vol. AC-14 No. 1, pp. 72-75.

WANG, S.H. 1970. "Design of precompensator for decoupling problem," Electron. Lett., 6, pp. 739-741.

WARNER, S. 1965. "Modern algebra," Prentice-Hall, Vol. 2.

WHITBECK, R.F. 1968. "A frequency domain approach to linear optimal control," Proc. JACC, pp. 726-737.

WHITTAKER, E.T. 1927. "Analytical dynamics of particles and rigid bodies," Cambridge University Press.

WIENER, N. 1949. "Extrapolation, interpolation and smoothing of stationary time series," MIT & Wiley, New York.

WILKIE, D.F. and PERKINS, W.R. 1969. "Essential parameters in sensitivity analysis," Automatica, Vol. 5, No. 2, pp. 191-197.

WILKIE, D.F. and PERKINS, W.R. 1969. "Generation of sensitivity functions for linear systems using low-order models," IEEE Trans. on Automatic Control, Vol. AC-14, pp. 123-130.

WILKINSON, J.H. 1965. "The algebraic eigenvalue problem," Clarendon Press, Oxford.

WILLIS, B.H. and BROCKETT, R.W. 1965. "The frequency domain solution of regulator problems," IEEE Trans., AC-10, pp. 262-267.

WINSOR, C.A. and ROY, R.J. 1970. "The application of specific optimal control to the design of desensitized model following control systems," IEEE Trans. on A.C. Vol. AC-15, No. 3.

WOLOVICH, W.A. 1969. "A frequency domain approach to the design and analysis of linear multivariable systems" NASA Technical Report C 104.

WOLOVICH, W.A. and FALB, P.L. 1969. "On the structure of multivariable systems" J. SIAM on Control, 7 (3), pp. 437-451.

WOLOVICH, W.A. and SHIRLEY, R.S. 1970. "A frequency domain approach to handling qualities design," Preprints of 1970 JACC.

WONHAM, W.M. and JOHNSON, C.D. 1964. "Optimal bang-bang control with quadratic index of performance," J. Basic Engineering, 86, pp. 107-115.

WONHAM, W.M. 1967. "On pole assignment in multi-input controllable linear systems," IEEE Trans., AC-12 (6), pp. 660-665.
WONHAM, W.M. 1967. "Optimal stationary control of a linear system with state-dependent noise," Journal of SIAM Control, Vol. 5, pp. 486-500.
WONHAM, W.M. 1970. "Dynamic observers - geometric theory," Trans. IEEE, AC-15, pp. 258-259.
WONHAM, W.M. and MORSE, A.S. 1970. "Decoupling and pole assignment in linear multivariable systems: A geometric approach," SIAM J. Control, 8, pp. 1-18.

YOULA, D.C. 1961. "On the factorization of rational matrices," IRE Trans. on Inf. Theo. Vol. IT-7, pp. 172-189.

ZADEH, L.A. 1950. "Correlation function and power spectrum in variable networks," Proc. IRE, Vol. 38, pp. 1342-1345.
ZADEH, L.A. 1961. "Time varying networks," Proc. IRE, Vol. 48, pp. 1488-1503.
ZADEH, L.A. and DESOER, C.A. 1963. "Linear system theory: The state space approach," McGraw-Hill, New York.
ZIEGLER, J.G. and NICHOLS, N.B. 1942. "Optimum settings for automatic controllers," Trans. ASME, 64, pp. 759-768.

ADDITIONAL REFERENCE

BELLETRUTTI, J.J. and MACFARLANE, A.G.J. 1971. "Characteristic loci techniques in multivariable control system design," to be published in Proceedings IEE.

M. Thoma

OPTIMAL MULTIVARIABLE CONTROL SYSTEMS THEORY: A SURVEY

OPTIMAL MULTIVARIABLE CONTROL SYSTEMS THEORY: A SURVEY

M. Thoma
Professor, Dr.-Ing.
Institut für Regelungstechnik,
Technische Universität Hannover
Hannover /Germany

ABSTRACT

The purpose of this paper is to present a tutorial summary of some important methods for optimal control. The first part deals with static optimization methods such as linear and nonlinear programming. The second part is devoted to the dynamic optimization problems such as maximum principle and dynamic programming for as well continuous as discrete lumped parameter systems, optimum control of stochastic and of distributed parameter systems as well as differential games.

1. INTRODUCTION

The problem of "optimization", where the word "optimum" may refer to either a maximum or minimum depending on the special situation, is as old as human being are living. It was always their desire to search for the "best" solution to the problem on hand.But just recently the field of optimization has reached a certain state of maturity. Why is it so? I like to point out some very close related reasons, why it took even for engineering problems such a long time. First of all it is by no means easy to give for a system a precise statement what is meant by "best" or "optimum". On the other hand practical problems do in general not have a unique solution; indeed, in most cases an infinite number of possible solutions exist. Therefore a performance criteria is needed in order to decide whether a solution is more close to the optimum than another one. Finally we want a strategy which leads to the optimal solution. These are very fundamental and typical problems, with which every engineer is confronted if he deals with optimization problems.

In this paper we consider optimal control problems. Their solution includes in general three important steps.

1. The set up of a mathematical model of the system to be optimized.

2. Formulation of a performance criteria (payoff function, value function, cost function).

3. Computation of the optimal control law (optimization methods).

The first two points are naturally very important for the solution of practical problems. Especially the choice of the "right" performance criteria, for example to find out which one is the most important one among several who have to be considered, causes for the engineer in many cases quite some problems. Still we are not dealing with questions like this but restrict ourselves to the third problem. A very important factor that has influenced this rapid growth of optimal control has been the development of practical, high-speed digital computer.

The enormous amount of publications in the field of systems optimization during the past decade indicates the intense research effort. Quite a part of this work is devoted to more or less pure mathematical questions such as: "Does an optimal control exist and if so, is the solution unique?" Here we are not dealing with these important questions because we like to restrict ourselves to the basic principles of optimal control from an engineering viewpoint. On the other hand in many cases an engineer is not very sure whether his mathematical model is a sufficiently good approximation of the considered problem at all. Therefore he has always to verify the results very critically. In other words an engineer is always confronted with the question whether the solution is a good description of the behaviour of the considered problem. In many practical cases the answers to the existence and uniqueness can to some extend be given by pure consideration of the practical problems (physical intuition). They therefore constitute not the most important fact in our consideration. We also pay no attention to the problem of sufficient conditions. A treatment of these more mathematical questions can be found in the books /1 - 10/. Also the survey papers of optimal control of deterministic systems /11/, of distributed parameter systems /12/, and of stochastic systems /13/ contain in their list of references a large amount of publications which are devoted to these problems. The square brackets refer to the representative but not at all exhaustive references on the end of this paper.

The scope of the paper is to give from an engineering point of view an <u>introduction</u> and a <u>short review</u> of some important optimization methods. Even for dissimular technical problems there exists at present some useful mathematical methods and procedures of optimal control which are very similar from a system synthesis viewpoint. In order to underline these similarities we characterize our survey due to the form of mathematical equations by which the plant is described and mathematical treatment of optimization but not by the area of interest and application. The survey paper /14/ is, for example, oriented towards aeronautical and astronautical applications. We are splitting the consideration up in two main parts, the <u>static</u> and the <u>dynamic</u> case. Other distinctions are the continuous- and discrete-time optimal control, lumped and distributed parameter systems, and the optimal control of deterministic and stochastic systems. Naturally there are always overlappings.

The optimal controller settings were one of the first main optimization problems with which the control engineers were engaged. In these parameter-optimal problems the structure of the controller (for example PID-controller) is choosen a priori. Then we attempt to determine the "optimum" (controller-) parameter in such a way that a performance index, e.g., the integral of a squared error is optimized. This problem leads in general to a classical maxima and minima problem. In this paper, we will deal especially as far as the dynamic case is concerned with a more general problem. In determining an optimal control strategy for a given system we do not make any prior assumptions or commitments that would fix the controller structure. We assume in this paper we have some information of the control object (plant) - such as its inputs, its outputs, and its dynamic structure and parameters. Our aim is to find an input function which will optimize the systems performance with respect to some cost criterion. Sometimes this function will depend on the output of the system; in that case we have a feedback solution.

Such a typical control problem begins with the following statement:

1. Let the mathematical model of a deterministic system be described by the vector differential equation

   $$\underline{\dot{x}} = \underline{f}(\underline{x},\underline{u},t) \tag{1.1}$$

   where $\underline{x}(t)$ is an n-vector representing the state of the system at time t, and $\underline{u}(t)$ is a r-vector representing the control input to the system at time t.
   In (1.1), $\underline{f}$ is a vector valued function of the state $\underline{x}(t)$ and the control $\underline{u}(t)$. For convenience we assume in most cases that the system is time-invariant, that is equation (1.1) does not depend on t explicitly; the same holds for the following performance function (1.2).

2. Let t_o be the initial and t_f the terminal time; in general, t_f may be specified a priori or not. The initial state $\underline{x}(t_o)$ is given and the final state $\underline{x}(t_f)$ is specified.

3. The types of performance index that one considers are functionals of the form

   $$J = \int_{t_o}^{t_f} F(\underline{x},\underline{u},t)\, dt \tag{1.2}$$

   wehre F is a prescribed scalar-valued cost function.

4. The state vector $\underline{x}(t)$ and/or the control vector $\underline{u}(t)$ are usually constraint. In other words, $\underline{u}$ and $\underline{x}$ must stay in the admissible regions Ω and X, respectively

$$\underline{u} \in \Omega \quad \text{for all } t, \tag{1.3}$$

$$\underline{x} \in X \quad \text{for all } t. \tag{1.4}$$

The general control problem consists in finding an optimal control $\underline{u}^*$ (and the terminal time t_f) which forces the dynamical system (1.1) from the given initial state $\underline{x}(t_o)$ to the specified final state $\underline{x}(t_f)$, subject to the constraints (1.3), and/or (1.4) and optimizes at the same time the performance index (1.2).

2. STATIC OPTIMIZATIONS

In contrast to the dynamic case where the systems are described by differential, difference, integral and functional-differential equations, are in the static case the systems characterized by algebraic equations. Important cases where the engineer has to deal with static optimization problems are on one hand systems which are by itself fully characterized by algebraic equations. Well known examples are the so called transportation problems (operations research). On the other hand we have problems where we restrict ourselves just to steady state of a dynamic system, e.g., electrical-utility plants.

Let us in the latter case assume that the system satisfies (1.1). If we like to keep the state at the constant, steady value $\underline{x} = \underline{x}^o$ the (dynamic) differential equation reduces to the (static) algebraic equation

$$\underline{f}(\underline{x}^o, \underline{u}^o) = \underline{0} \tag{2.1}$$

from which we may solve for the constant steady control force $\underline{u}^o$ that will keep the system in the state in question. With both $\underline{x}$ and $\underline{u}$ reduced to constants, we conclude that the rate of change of the payoff function (1.2) also turns to the constant value

$$\left(\frac{dJ}{dt}\right)^o = F(\underline{x}^o, \underline{u}^o). \tag{2.2}$$

We conclude that the cost index grows at a constant rate, and we therefore can optimize the process by optimizing this rate.

2.1. Linear Programming

One of the most simple class of problems is if the static optimization can be formulated as well by a linear algebraic equation as by a linear cost function. In this case we speak of a linear programming problem /15 - 17/. The linear optimization problem can always be expressed in the following way. Find nonnegative x_i minimizing the linear cost index

$z = \sum_{i=1}^{n} c_i x_i$ subject to the linear inequality constraints

$$\sum_{i=1}^{n} a_{ji} x_i \leq b_j \quad (j= 1,2,...,m).$$

An equivalent analytic formulation is

$\underline{c}^T \underline{x} = z = \text{Min!}$	(cost index),	(2.3)
$\underline{A}\ \underline{x} \leq \underline{b}$	(inequality constraints),	(2.4)
$\underline{x} \geq \underline{0}$	(nonegativity conditions).	(2.5)

By introducing of (nonegative) slack variables the inequality constraints (2.4) can always be transformed to equality constraints. In case of a two-dimensional problem ($\underline{x} \in R^2$) a simple graphical interpretation can be given. The cost index (2.3) is a family of parallel straight lines which move further away from the origin of the x_1,x_2-plane as z is increased through positive values. The constraints (2.4) and the conditions (2.5) form a polygonal area. As the cost function is for every fixed z a straight line it is obvious that the optimum value must be located at the periphery of this area, quite probably in the corner points (extreme points). It can be proved that the optimal solution must lie on at least one of the corner points. This is also true in a high-dimensional case. Where the cost function (2.3) forms a hyperplane and the constraints (2.4) together with the conditions (2.5) form a polyhedron (simplex). Linear programming implies therefore a straight forward search by exchanging the corners, that is moving from one basic feasible solution to another.

For higher dimensional problems the process would even with the use of high-speed digital computers in general be uneconomical because of too much computing time needed. A well known more economic procedure is the simplex method /15,18/. At each iteration the simplex algorithm moves from one basic feasible solution to another and at the same time improves the value of the cost function.

The usual amount of computational effort required for solving a problem growth much faster then the size of the problem. One of the main difficulties to find a solution for large-scale systems is that it exceeds the storage capacity and/or too much computer time is needed. For these reasons methods have been developed for decomposing the "large" problem to a number of "small" problems in such a manner that they can be independently optimized. The interaction between the subsystems are selected by the decomposition principle for linear programming /18 - 20/ such that after optimization of the subsystems the total system is optimized.

If the cost function (2.3) and/or the restrictions (2.4) and (2.5) can depend on stochastic parameters then we have the special problems of(linear) stochastic programming or (linear) programming under uncertainty. Methods were developed in order to find for example the optimum of the expectation and/or variance of the cost functions /21-23/.

2.2. Nonlinear Programming

A suitable extension of the above concepts to nonlinear optimization problems leads to the following formulation.

Minimize the nonlinear performance index

$$F(\underline{x}) = \text{Min!} \tag{2.6}$$

subject to the nonlinear inequality constraints

$$f_j(\underline{x}) \leq \underline{0} \qquad (j = 1,2,\ldots m) \tag{2.7}$$

and the nonnegativity conditions

$$\underline{x} \geq \underline{0}\,. \tag{2.8}$$

F and all f_j are assumed to be continuous real-valued functions over the set of $\underline{x} \in R^n$.

A rigorous analysis of this general nonlinear optimization problem has remained rather intractable. In most cases the solution is based on more or less suitable optimum seeking methods which in general leads not to a global but to a local optimum. In this case where $F(\underline{x})$ and all $f_j(\underline{x})$ are convex functions effective methods are existing. This convex programming problem has received particular attention /18,24-27/. One reason is because every local optimum is also a global optimum. Analogous to linear programming this is obvious from geometric considerations in R^2, but the optimum value must not be located at the periphery of the convex domain. A central part in convex programming plays the Kuhn-Tucker-Theorem a necessary and sufficient condition that a certain $\underline{x}^*$ be a solution of the nonlinear convex programming problem. It is a generalization of the classical Lagrange multipliers method and equivalent to the saddle point problem.

We define the Lagrangian function

$$\underline{\phi}(\underline{x},\underline{\lambda}) = F(\underline{x}) + \underline{\lambda}^T \underline{f}(\underline{x}) \tag{2.9}$$

where the m-vector $\underline{\lambda}$ is the Lagrange multiplier.

Necessary and sufficient conditions that $\underline{x}^*$ be a solution of the convex programming problem
Min $\{F(\underline{x}) \mid \underline{x} \geq \underline{0};\ \underline{f}(\underline{x}) \leq \underline{0}\}$ is that there exists a vector $\underline{\lambda}^*$ such that

$$\underline{x}^* \geq \underline{0}, \quad \underline{\lambda}^* \geq \underline{0} \tag{2.10}$$

and

$$\phi(\underline{x}^*,\underline{\lambda}) \leq \phi(\underline{x}^*,\underline{\lambda}^*) \leq \phi(\underline{x},\underline{\lambda}^*) \tag{2.11}$$

for all $\underline{x} \geq \underline{0}$ and $\underline{\lambda} \geq \underline{0}$. (2.12)

Because of (2.11) $\underline{x}^*$ and $\underline{\lambda}^*$ constitute a saddle point for $\phi(\underline{x},\underline{\lambda})$ a global condition for the Lagrangian function.

Of the convex functions $F(\underline{x})$ and $f_j(\underline{x})$ are continuously differentiable then the saddle point condition can be replaced by the equivalent local conditions

$$\phi_x(\underline{x}^*,\underline{\lambda}^*) \geq 0, \qquad \underline{x}^{*T}\phi_x(\underline{x}^*,\underline{\lambda}^*) = 0 \tag{2.13}$$

$$\phi_u(\underline{x}^*,\underline{\lambda}^*) \leq 0, \qquad \underline{u}^{*T}\phi_u(\underline{x}^*,\underline{\lambda}^*) = 0 \tag{2.14}$$

$$\underline{x}^* \geq \underline{0} \quad \text{and} \quad \underline{u}^* \geq \underline{0} \tag{2.15}$$

were ϕ_x and ϕ_u are the gradient vectors with regard to $\underline{x}$ and $\underline{u}$ respectively. The local conditions are usually easier to apply. Very effective methods have been developed for special convex programming forms, such as the quadratic programming problem in which the performance function is quadratic but the constraints are linear, Min $\{\underline{c}^T\underline{x} + \underline{x}^T\underline{Q}\,\underline{x} \mid \underline{A}\,\underline{x} \leq \underline{b},\ \underline{x} \geq \underline{0}\}$ with $\underline{Q}$ a symmetric n,n positive definite matrix. For example the method of Wolfe /28/, a modification of the simplex method, provides a solution of the Kuhn-Tucker-conditions and therefore also of the quadratic optimization problem, if a solution exists at all. A survey of several other computational methods in quadratic programming are given in /18,24-27, 29/.

The arguments we made according large-scale systems for linear programming problems also hold for convex programming problem. By decomposition a class of results can be obtained, which are useful for large problems /30/. The principles of decomposition and multilevel optimization are very close related. The subsystems constitute the first level. A second level controller then manipulates certain parameters in the first level such that after repeated optimization of the subsystems, the total system is optimized /31/.

Important techniques for the solution of convex and even nonconvex programming problems are the search methods, e.g., the class of gradient methods. They all involve some form of comparison, e.g., a function $f(\underline{x})$ is evaluated at various allowed values of $\underline{x}$, and certain values of $f(\underline{x})$ are compared in order to find an estimate of the optimum $f(\underline{x})$. Because of limited space we do not give a comparison of the many different search techniques /32 - 34/. For a given particular optimization problem a searcher's prime concern is to utilize a search technique which not only solves the problem, but also solves it efficiently. The efficiency of a given search technique is affected by certain global and local properties of functions. For functions (of one and more variables) some of them are more appropriately used on analytically given functions while others are more useful when data are obtained experimentally. The set of search techniques may also be subdivided in the categories of sequential and non-sequential techniques. Nonsequential search methods are generally for higher-dimensional search unefficient, but are useful under important special conditions. For constraint optimization problems such as the nonlinear programming, penalty functions are used in order to transform the constraint problem into a sequence of unconstraint optimizations /35/. Although many search techniques are directly applicable to nonlinear search problems, piecewise linear approximations are still much used; linear programming is applied over a restricted range of the variables, and the linear model is updated periodically.

The above mathematical-model approach which represents an openloop optimization can only be used when the process is well defined, its parameters generally known and the effects of disturbances negligible. But the search methods are also of great practical interest if a sufficiently accurate mathematical model is unobtainable. It is still possible under certain conditions to keep a control process at its static optimum by "experimenting" with the system in real time. This optimization by experimentation which represents a closed-loop optimization is to a great extent based on "hill-climbing techniques /36 - 38/". Adaptive control and computer process control are other closed-loop techniques which are very close related to optimization techniques /38 - 40/.

An important application finds the mathematical (linear) programming in the theory of games; e.g. multiple-stage quantitative games. In the class of two-player, zero-sum games, one player seeks to minimize while the other player seeks to maximize a payoff function which is additive over the stages over the play. Optimality is expressed by a saddle point condition (Min-Max problem); the corresponding value of the payoff function is called the value of the game /26, 41 - 44/.

Finally we like to point out that as well for static as for dynamic optimization problems the question of sensitivity is of great practical interest, because the system parameters can only be selected within certain tolerances. The immediate practical questions to be answered are: How is the value of the payoff function affected if a suboptimal feasible solution is used rather than the optimal solution? And how does the nature of the optimal solutions change when (parameter) changes are made in the constraints and payoff function?

3. DYNAMIC OPTIMIZATION

Because automatic control mainly involves dynamical systems, dynamic optimization problems are of very great importance. The dynamic optimization may to some extent be considered as a generalization of the static optimization and there exists quite a number of similarities. Whereas in the static case the performance criteria (2.3) or (2.6) is a function, in the dynamic case it is a functional of the form (1.2) that depends on the entire histories of $\underline{x}(t)$ and $\underline{u}(t)$ over $t_o \leq t \leq t_f$ and in the general case on the initial time t_o, initial state $\underline{x}(t_o)$, terminal time t_f as well as terminal state $\underline{x}(t_f)$. As we already mentioned a good performance is difficult to define. But we like to enumerate some cases of typical classes of optimal control problems.

1. Time-optimal control

$$J = \int_{t_o}^{t_f} dt = t_f - t_o \qquad (t_f \text{ free}). \qquad (3.1)$$

2. Fuel-optimal problem

$$J = \int_{t_o}^{t_f} \sum_{i=1}^{r} |u_i(t)| \, dt \qquad (3.2)$$

3. Minimum integral-square error problem

$$J = \int_{t_o}^{t_f} (\underline{x}^T \underline{Q} \, \underline{x}) \, dt. \qquad (3.3)$$

4. Minimum energy problem

$$J = \int_{t_o}^{t_f} (\underline{u}^T \underline{R}\ \underline{u})\ dt \qquad (3.4)$$

5. Terminal (or final value) optimal control problem

$$J = \Theta\left[\ \underline{x}(t_f), t_f\ \right] \qquad (\Theta \text{ scalar-valued function}). \qquad (3.5)$$

Frequently it is more practical and even necessary to combine some of the above simple performance indices, e.g.

$$J = \frac{1}{2}\ \underline{x}^T(t_f)\ \underline{S}\ \underline{x}(t_f) + \frac{1}{2} \int_{t_o}^{t_f} (\underline{x}^T \underline{Q}\ \underline{x} + \underline{u}^T \underline{R}\ \underline{u})\ dt \qquad (3.6)$$

which represents a combination of (3.3), (3.4) and a special case of (3.5).

Often some (or all) of the end states are not specified. In that case the transversality conditions supplies the "missing" boundary condition.

The above differentiation refers to the application of optimal control to different practical problems. Without a thorough understanding of the physical principles upon which a given problem solution depends the application of optimization principles is of dubious value. For example all practical problems are usually constraint in some kind of a way. If we thus consider a time optimal control of a linear time invariant system where all the components of the control vector $\underline{u}(t)$ are constraint in magnitude by the relaticns $|\ u_i(t)\ | \leq 1$, then the control is (under certain assumptions) "bang-bang" and switches between the constraints. Although the framework of "calculus of variations" is the basis of optimal control its direct application is limited to a rather small class of practical problems because methods for treating constraint relationships - cf. eqs.(1.3) and (1.4) - have been neglected almost entirely. A real progress in this field started with the important works of the "Maximum Principle" /45/ and Dynamic Programming" /46/. For additional reading, standard works are available at various level of depth /47-60/.

3.1. The Maximum Principle

The maximum principle which can be viewed as an extension and application of the classical calculus of variations to the optimal control problem was originally developed for continuous-time problems. Later it has been extended to discrete-time problems.

First we consider that on pages 3 and 4 formulated typical optimum control problem (No. 1 to 4). Whereby we assume that eqs. (1.1) and (1.2) do not depend on t explicitly. In addition let the "target set" S be a surface, or manifold, or point out in the n-dimensional state space. As a part of the boundary conditions for the optimization problem, one may demand that the state vector at the terminal time t_f belong to the target set, i.e.,

$$\underline{x}(t_f) \in S. \tag{3.7}$$

The precise statement of the optimization problem is then as follows: Given the system

$$\dot{\underline{x}} = \underline{f}\left[\underline{x}(t),\ u(t)\right] \tag{3.8}$$

the boundary conditions $\underline{x}(t_o) = \underline{x}_o$, the control constraint set Ω , the target set S , and the cost functional

$$J(\underline{u}) = \int_{t_o}^{t_f} F\left[\underline{x}(t),\underline{u}(t)\right] dt. \tag{3.9}$$

Then find the control that satisfies the constraint $\underline{u}(t) \in \Omega$, transfers the state of the system (3.8) from $\underline{x}(t_o) = \underline{x}_o$ to $\underline{x}(t_f)$ so that $\underline{x}(t_f) \in S$, and minimizes the cost functional (3.9).

The above formulation is rather a minimization than a maximization problem subject to the constraint in form of the vector differential equations of the controlled system. We therefore introduce certain variables (functions) which are analogous to the Lagrange multipliers, cf. eq. (2.9). These Lagrange multipliers are often called "costate variables", and are denoted by $p_1(t)$, $p_2(t)$, $\ldots, p_n(t)$, where n is the dimension of the state vector $\underline{x}(t)$. The costate variables form the costate vector $\underline{p}(t)$. The scalar-valued function

$$H\left[\underline{x}(t),\underline{\mathbf{p}}(t),\underline{u}(t)\right] = F\left[\underline{x}(t),\underline{u}(t)\right] + \underline{\mathbf{p}}^T(t)\underline{f}\left[\underline{x}(t),\underline{u}(t)\right] \tag{3.10}$$

is called the "Hamiltonian function" or just "Hamiltonian". Suppose that an optimal control $\underline{u}(t)$ exists for $t_o \leq t \leq t_f$. For simplicity, assume that the terminal time t_f is specified a priori. Let $\underline{x}^*(t)$, $t_o \leq t \leq t_f$, denote the solution of the differential equation (3.8) when the optimal control $\underline{u}^*(t)$ is applied. It is common to refer to $\underline{x}^*(t)$ as the optimal trajectory generated by the control $\underline{u}^*(t)$. Basically the minimum (maximum) principle provides a set of local necessary conditions for optimality. Its statement is as follows: /11,47/:

Suppose that $\underline{u}^*(t)$ is the optimal control and that $\underline{x}^*(t)$ is the generated optimal trajectory. Then, corresponding to $\underline{u}^*(t)$ and $\underline{x}^*(t)$, there exists a costate vector $\underline{p}^*(t)$ such that the following relations hold.

1. Canonical Equations

$$\frac{d}{dt} x_i^*(t) = \left. \frac{\partial H}{\partial p_i(t)} \right|_* \tag{3.11}$$

$$\frac{d}{dt} p_i^*(t) = \left. \frac{\partial H}{\partial x_i(t)} \right|_* \tag{3.12}$$

2. Boundary Conditions

$$\underline{x}^*(t_o) = \underline{x}_o \tag{3.13}$$

$$\underline{x}^*(t_f) \in S \tag{3.14}$$

$$\underline{p}^*(t_f) \text{ normal to } S \text{ at } \underline{x}^*(t_f). \tag{3.15}$$

3. Minimization of the Hamiltonian

$$H\left[\underline{x}^*(t)\underline{p}^*(t),\underline{u}^*(t)\right] \leq H\left[\underline{x}^*(t),\underline{p}^*(t),\underline{u}(t)\right] \tag{3.16}$$

for every t in $t_o \leq t \leq t_f$, and all $\underline{u}(t) \in \Omega$.

When one uses the minimum principle in order to calculate the optimal control $\underline{u}^*(t)$ the following fundamental steps are necessary. One starts out from the relation (3.16) in order to deduce in general a relation of the form

$$\underline{u}^*(t) = \underline{g}\left[\underline{x}^*(t),\underline{p}^*(t)\right] \tag{3.17}$$

If $\underline{x}^*(t)$ and $\underline{p}^*(t)$ uniquely specify $\underline{u}^*(t)$, then one deals with the socalled "normal" problem. However, if relation (3.16) is over a finite time-intervall satisfied even though $\underline{u}^*(t) \neq \underline{u}(t)$ then we speak of a "singular" problem. Consider a normal problem. If we substitute the relations (3.17) into the 2n canonical differential equations (3.11) and (3.12), they depend only on $\underline{x}^*(t)$ and $\underline{p}^*(t)$. We have thus 2n differential equations for the determination of the required 2n functions $\underline{x}^*(t)$ and $\underline{p}^*(t)$. The solution of the optimal state vector (trajectory) $\underline{x}^*(t)$ and the optimal costate vector $\underline{p}^*(t)$ leads to a two point boundary value problem. The inital and final value conditions or/and the transversality conditions and if the terminal time t_f is free one additional equations, e.g., $H\left[\underline{x}^*(t),\underline{p}^*(t),\underline{u}^*(t)\right] = 0$ for $t_0 \leq t \leq t_f$, provides at least theoretically the necessary conditions. After one has found the optimal state and costate vectors one gets the optimal control vector $\underline{u}^*(t)$ by substitution of $\underline{x}^*(t)$ and $\underline{p}^*(t)$ in equ. (3.17).

It is obvious that only relative simple problems in optimal control can be solved analytically. This is the reason why numerical solutions take on added importance. We come back to this point in section 3.5.

3.2. Discrete Maximum Principle

From a computational aspect the discrete maximum principle has some advantages over the continuous maximum principle because it is in a form directly suitable for digital computation. One could even think to discretize the problem from the start and deal throughout with difference, rather than differential equations. The approximation of a continuous time system by its discrete analog has its merits. But from an engineering viewpoint it is not only the computational aspect why one deals with optimal control problems. The intuitive insight into the structure of an optimal system which is at least as important as the computational aspect generally is reduced by discretization. The optimization of sampled-data systems at sampling instances is another problem for application of the discrete maximum principle. The same holds for systems which are discrete by nature.

For quite some time there has been some confusion in the literatur about the assumptions and validity of the discret maximum principle. A few years ago a proof of the necessary conditions was given. The proof is based upon assumptions that guarenteed that the sets of realable states were convex (directionally convex). /61,62/. It can, however, be shown that, in the approximation of the continuous-time systems by its discrete analog, convexity is always assured. However, when the system is discrete in nature, this can no longer be assured. For an excellent discussion on these conditions see /11/.

As in the time-continuous case the necessary conditions will be stated according /11/ in terms of the minimum principle. Under the above assumptions the statement of the discrete-time optimal problem is as follows:

Given the dynamical system which is described by the vector difference equation

$$\underline{x}_{i+1} - \underline{x}_i = \underline{f}_i \left[\underline{x}_i, \underline{u}_i \right] \quad (i = 0,1,2,\dots,N-1) \tag{3.18}$$

where $\underline{x}_i$ is the value of the state vector at the i-th instant of time, $\underline{u}_i$ the value of the control vector at the i-th instant of time, and $\underline{f}_i$ is a vector-valued function of the state $\underline{x}_i$ and of the control $\underline{u}_i$.
Given the control constraint

$$\underline{u}_i \in \Omega \qquad \text{for all } i = 0,1,2,\dots,N-1 \tag{3.19}$$

and the cost functional

$$J = \sum_{i=0}^{N-1} F_i(\underline{x}_i, \underline{u}_i) \tag{3.20}$$

where $F_i(\underline{x}_i, \underline{u}_i)$ is a scalar-valued function of $\underline{x}_i, \underline{u}_i$. It is assumed that the following boundary conditions are specified:

$$\text{At } i = 0: \quad \underline{x}_o = \underline{\alpha}, \text{ at } i = N: \ \underline{x}_N \in S \tag{3.21}$$

where S is a target set in the state space.
Find the sequence of optimal control $\underline{u}_o^*, \underline{u}_1^*, \dots, \underline{u}_{N-1}^*$ satisfying the constraint (3.19) such that the generated state sequence $\underline{x}_o^*, \underline{x}_1^*, \dots, \underline{x}_{N-1}^*$ has the property that $\underline{x}_N^* \in S$, and such that the cost functional (3.20) attains its minimum value.

As it was the case in the continuous-time case, we introduce a sequence of costate vectors (Lagrange multipliers) denoted by $\underline{p}_i$. The Hamiltonian is in this case defined by the relation

$$H_i(\underline{x}_i, \underline{p}_{i+1}, \underline{u}_i) = F_i(\underline{x}_i, \underline{u}_i) + \underline{p}_{i+1}^T \underline{f}_i(\underline{x}_i, \underline{u}_i) \tag{3.22}$$

for i = 0,1,...,N-1. The statement of the discrete minimum principle is as follows /11/:

Suppose that $\underline{u}^*_o, u^*_1, \dots, u^*_{N-1}$ is the optimal control sequence. Let $\underline{x}_o^*, \underline{x}_1^*, \dots, \underline{x}_N^*$ denote the generated optimal state sequence. Then there is a corresponding costate sequence $\underline{p}_o^*, \underline{p}_1^*, \dots, \underline{p}_N^*$ such that the following relations hold.

1. Canonical Difference Equations:

$$\underline{x}_{i+1}^* - \underline{x}_i^* = \left. \frac{\partial H_i}{\partial \underline{p}_{i+1}} \right|_* \tag{3.23}$$

$$\underline{p}_{i+1}^* - \underline{p}_i^* = - \left. \frac{\partial H_i}{\partial \underline{x}_i} \right|_* \tag{3.24}$$

2. Boundary Conditions

$$\underline{x}_o^* = \underline{\alpha} , \quad (3.25)$$

$$\underline{x}_N^* \in S, \quad (3.26)$$

$$\underline{p}_N^* \text{ normal to S at } \underline{x}_N^*. \quad (3.27)$$

3. Minimization of the Hamiltonian

$$H_i(\underline{x}_i^*, \underline{p}_{i+1}^*, \underline{u}_i^*) \leq H_i(\underline{x}_i^*, \underline{p}_{i+1}^*, \underline{u}_i) \quad (3.28)$$

for all $\underline{u}_i \in \Omega$ and all $i = 0,1,\ldots,N-1$.

The analogy of the continuous and discrete maximum (minimum) principle is obvious.

3.3. The Linear Regulator Problem

In general, the application of the maximum principles in order to generate an optimal control $\underline{u}^*(t)$ leads to an "open loop" control. If the optimal control $\underline{u}^*$ can be expressed as a function of the state $\underline{x}(t)$, e.g. $\underline{u}^*(t) = g\,[\,\underline{x}^*(t),t\,]$, then we have a solution in a feedback form. The optimal control of the linear plant

$$\dot{\underline{x}} = \underline{A}(t)\ \underline{x}(t) + \underline{B}(t)\ \underline{u}(t)\ ,(\underline{x}(t) \in R^n) \quad (3.29)$$

with respect to the quadratic performance index (3.6) leads to a feedback control law. We assume that the symmetric matrix $\underline{S}$ is positive semidefinite, and the timevariable symmetric matrices $\underline{Q}(t)$a. $\underline{R}(t)$ are positive definite. Then the optimal control $\underline{u}^*(t)$ which minimizes the cost functional (3.6) exists, is unique and is given by the linear control law,

$$\underline{u}^*(t) = -\underline{R}^{-1}(t)\ \underline{B}^T(t)\ \underline{P}(t)\ \underline{x}^*(t) = -\underline{K}(t)\underline{x}(t) \quad (3.30)$$

where $\underline{P}(t)$ is a solution of matrix Riccati differential equation

$$\frac{d}{dt}\ \underline{P}(t) = -\underline{P}(t)\underline{A}(t) - \underline{A}^T(t)\underline{P}(t) + \underline{P}(t)\underline{B}(t)\underline{R}^{-1}(t)\underline{B}^T(t)\underline{P}(t) - \underline{Q}(t) \quad (3.31)$$

subject to the boundary condition

$$\underline{P}(t_f) = \underline{S} \quad (3.32)$$

The n,n-matrix $\underline{P}(t)$ is symmetric, positiv definite and has n(u+1)/2 different terms. This result can be used to

design optimal linear feedback systems, c.f. /47,63/.
If as well the matrices $\underline{A}$ and $\underline{B}$ in (3.29) as the matrices $\underline{Q}$ and $\underline{R}$ in (3.6) are constant, if $\underline{S} = \underline{O}$ and $t_f \rightarrow \infty$, then the optimal feedback systems turns out to be linear and time invariant. In this case, $\underline{K}$ in the linear optimal control law (3.30) is constant, and $\underline{P}$ is a symmetric and positive definite solution of the constant algebraic matrix equation - (steady state solution of eq. (3.31)) -

$$- \underline{P}\,\underline{A} - \underline{A}^T\underline{P} + \underline{P}\,\underline{B}\,\underline{R}^{-1}\underline{B}^T\underline{P} - \underline{Q} = \underline{O}. \qquad (3.33)$$

It is important to note that all components of the state vector must be accessible. If this is not the case it is possible to construct a "state estimator" /64,65/.

Continuous time dynamic programming has as we shall see in the next section the property that the optimal control function $\underline{u}^*$ in usually in the feedback form. However, it requires that we must solve instead of a set of 2n ordinary differential equations (if we use calculus of variations or the maximum principle) a single partial differential equation with one-point boundary (initial) conditions. In case of the above stated linear problem with a quadratic cost function the partial differential equation of dynamic programming eq. (3.40) or eq.(3.41) is direct solvable and one obtains the matrix Riccati equation (3.31) /48,58/.

3.4. Dynamic Programming

The dynamic programming method which is based on the principle of optimality was originally developed for discrete(-time) problems, and later on it was extended to continuous time processes. We also begin with discrete problems which can be formulated as a "multistage decision process". The phrase implies a problem which can be viewed as a succession of decision problems, each one building on the last, until the problem is solved.
We consider process state to be transformed by the decision made at each stage. Thus with dynamic programming problems there are associated the categories of state and decision variables. The state variable is denoted by "state" vector $\underline{x}$ and the decision variables which may also be refered to as control variables or policy variables by u_i. The "optimal policy" is that decision sequence or decision rule providing a maximum profit (maximum return) or minimum cost.

We start with a one-stage problem: If one likes to transform the state of a system from $\underline{x}^1$ to $\underline{x}^2$, that is $\underline{x}^2 = g(\underline{x}^1, u_1)$ then this cost after the first-stage is $R_1 = r(\underline{x}^1, u_1)$. The decision u_1 has to be made in such a manner that we get after the first stage the minimum cost $I_1(\underline{x}^1) = \min_{u_1} r(\underline{x}^1, u_1)$.

We proceed with a two-stage problem: First we transform with the relation $\underline{x}^2 = g(\underline{x}^1,u_1)$ the system from $\underline{x}^1$ to $\underline{x}^2$ and then with the relation $\underline{x}^3 = g(\underline{x}^2,u_2)$ from $\underline{x}^2$ to $\underline{x}^3$. The total cost are $R_2 = r(\underline{x}^1,u_1) + r(\underline{x}^2,u_2)$. Now a sequence of decisions u_1,u_2 have to be made in order to minimize after the second stage the total cost

$$I(\underline{x}^1) = \min_{u_1,u_2} \left[r(\underline{x}^1,u_1) + r(\underline{x}^2,u_2)\right].$$

By generalization the maximum total cost after stage N are

$$I_N(\underline{x}^1) = \min_{u_i} \sum_{i=1}^{N} r(\underline{x}^i,u_i) . \qquad (3.34)$$

Now the principle of optimality holds, which can be stated in the form:
An optimal policy has the property, that whatever the initial state and the initial decision are, the remaining decisions must constitute an optimal policy with regard to the state resulting from the first decision.
If we apply the principle of optimality to the N-stage process we can write the total cost

$$R_n = r(\underline{x}^1,u_1) + I_{N-1}\left[g(\underline{x}^1,u_1)\right] \qquad (3.35)$$

and appropriate for the minimal total cost

$$I_N(\underline{x}^1) = \min_{u_i}\left\{r(\underline{x}^1,u_1) + I_{N-1}\left[g(\underline{x}^1,u_1)\right]\right\} . \qquad (3.36)$$

The recurrence relation (3.35) expresses the step-wise nature of dynamic programming which allows each step (e.g. time-interval) to be optimized independend of all the other steps. It reduces the N-stage process to a sequence of N one-stage decision processes.

This effect reduces the computational problem by orders of magnitude, and transforms the problem from simultaneously solving to sequentially solving for optimal control variables.

Let us now apply the principle of optimality to the optimal control of continuous-time systems. We consider the performance index

$$J^*(\underline{x}_o,t_o) = \min_{\substack{\underline{u}(t)\in\Omega \\ t_o\le t\le t_f}} \int_{t_o}^{t_f} F(\underline{x},\underline{u},t)\,dt \qquad (3.37)$$

which can be written if we devide the time interval $t_o \leq t \leq t_f$ in the "first step" $t_o \leq t \leq t_o + \Delta t$ and the remaining "infinite step" $t_o + \Delta t < t \leq t_f$

$$J^*(\underline{x}_o, t_o) = \min_{\substack{\underline{u}(t) \in \Omega \\ t_o \leq t \leq t_f}} \left[\int_{t_o}^{t_o+\Delta t} F(\underline{x},\underline{u},t)dt + \int_{t_o+\Delta t}^{t_f} F(\underline{x},\underline{u},t)dt \right] . \tag{3.38}$$

Equation (3.38) can then be considered as a multistage decision process with a suitable form for the application of the principle of optimality: The optimal control function $\underline{u}^*(t)$ over $t_o \leq t \leq t_f$ has the property, that for any admissible Δt regardless of what the value of $\underline{u}^*(t)$ may be over $t_o \leq t \leq t_o + \Delta t$ and hence regardless of the value of $\underline{x}(t_o + \Delta t)$, it must still remain optimal with respect to the state $\underline{x}(t_o + \Delta t)$ over the time interval $t_o + \Delta t < t \leq t_f$. Applying the principle of optimality, eq. (3.38) can be converted into the form

$$J^*(\underline{x}_o, t_o) = \min_{\substack{\underline{u}(t) \in \Omega \\ t_o \leq t \leq t_o + \Delta t}} \left[\int_{t_o}^{t_o+\Delta t} F(\underline{x},\underline{u},t)dt + J^*\left(\underline{x}(t_o+\Delta t), t_o+\Delta t\right) \right] . \tag{3.39}$$

Expanding the integral of (3.39) in a Taylor series ab out $t=t_o$ and taking the limit $\Delta t \rightarrow 0$ provides

$$\frac{\partial J^*}{\partial t} + \min_{\underline{u}} \left[F(\underline{x},\underline{u},t) + \left(\frac{\partial J^*}{\partial \underline{x}}\right)^T \underline{f}(\underline{x},\underline{u},t) \right] = 0 \tag{3.40}$$

which is called Bellman's functional equation. With the Hamiltonian eq. (3.40) we may rewrite the functional eq. (3.40) as

$$\frac{\partial J^*}{\partial t} + H^*(\underline{x}, \frac{\partial J^*}{\partial \underline{x}}, t) \tag{3.41}$$

with

$$H^*(\underline{x}, \frac{\partial J^*}{\partial \underline{x}}, t) = H(\underline{x}, \frac{\partial J^*}{\partial \underline{x}}, \underline{u}, t)\Big|_{\underline{u}=\underline{u}^*} = \min_{\underline{u}} H(\underline{x}, \frac{\partial J^*}{\partial \underline{x}}, \underline{u}, t) .$$

Equation (3.41) is called the "Hamilton-Jacobi" partial differential equation and proves a necessary condition for the optimal solution.

Because the integral in eq.(3.39) vanishes for $t_o = t_f$ we get $J^*(\underline{x}, t_f)$ which is an initial condition (for the terminal time).

Needless to say, the Hamilton-Jacobi equation cannot be easily solved in general. However, when it can, $\underline{u}^*(t)$ is determined as a function of $\underline{x}(t)$ or in other words, we find a feedback solution which is highly desirable from an engineering viewpoint. The Hamilton-Jacobi partial differential equation is equivalent to the Bellman functional equation /55,56/. For application of the dynamic programming see /66,67/.

3.5. Computational Methods

The application of variational calculus, including the maximum principle, reduces the optimal control problem to two-point boundary-value differential or difference equations. Only for relatively simple problems it is, in general, possible to find an expression for the optimal control. Therefore, we must employ numerical methods for the optimal solutions. However, the amount of numerical computation required for even relative simple problems is very high and can not be done by hand. That's why variational methods found very little use in engineering and applied science until the development of economical, high-speed computers. At present time, there are many numerical methods for the solution of optimal problems; most of them are of iterative procedures. Because up to now there is no general method it is difficult for the practicing engineer to choose for a certain problem the "right" computing algorithm, e.g., in the sense that it has a fast rate of convergence.

One (direct) method which circumvents the difficulties associated with two-point boundary-value problems is the discrete dynamic programming. Its biggest disadvantage is the requirement for an enormous amount of computer memory even for moderately complicated problems. The polynomical approximation method /68/ and the state increment dynamic programming /67/ are two methods with reduced computational requirements.

The most common used numerical methods are the iterative computational algorithms. Let use illustrate some of the different ideas that are beeing used for the evaluation of optimal control problem using iterative techniques. All the proposed iteration procedures which are usually divided into direct and indirect methods, use "successive linearization". They have in common that by iterative algorithms some kind of a first approximation (estimation) is improved until some in the optimal sense specified conditions are satisfied (successive approximation).

A very popular direct method is the gradient method or method of steepest decent. They are characterized by iterative algorithms for improving estimates of the control vector $\underline{u}_o(t)$ so as to come closer to satisfying the boundary conditions (3.13) through (3.15) and the inequality (3.16) or until the cost functional is minimized. The gradient methods have the ability to generate successively improved trajectories with very poor starting values. However they tend to converge slowly as the optimal trajectory is approached. The gradient methods are based primarily on first order theory in a way of finding "first order" effects of controls upon terminal constraints and the cost function.

To improve the convergence near the optimal trajectory, second-order terms can be added. Due to the simularity of these terms in second variations, they are then called second-variation methods. They do accelerate convergence of gradient method but they require good initial estimates of the initial control and trajectory. Of course, the gradient computation may serve to generate initial iterations for the second-variation method, or the two procedures may be combined.

The quasilinearization method or as it is often called the generalized Newton-Raphson method is considered by most to be an indirect method. It nevertheless shares many simularities with the second-variation method, the latter being a direct method. In this method one commonly guesses trajectories $\underline{x}(t)$ and $\underline{p}_o(t)$ which will meet the boundary conditions (3.13) through (3.15). The nominal vector, $\underline{u}_o(t)$, is then determined by use of the inequality (3.16). The guesses $\underline{x}_o(t)$ and $\underline{p}_o(t)$ will not, in general, satisfy the systems differential equations (3.8). One then linearizes the differential equations about the initial guesses $\underline{x}_o(t)$ and $\underline{p}_o(t)$. The linearized equations are solved for a new set of trajectories which still meet the boundary conditions. The procedure is repeated until the trajectories approach the solution of the systems differential equations (3.8). This method converges near the optimal solution just as rapidly as the second-variations method. Besides this, it is sometimes helpful to know that it is often easier to guess a trajectory $\underline{x}_o(t)$ then a control $\underline{u}_o(t)$. Direct integration methods (neighbouring extremal algorithms) are characterized by iterative algorithms for improving estimates of the unspecified initial (or terminal) conditions so as to satisfy the specified terminal (or initial) conditions.

The main difficulty with these method is, finding a first estimate of the unspecified conditions at the other end. Unfortunately, because of the fact that the normal optimal control is generally very sensitive to small changes in the unspecified boundary conditions, these methods have not been completely successful.

As stated before, there is not a categorically "best" numerical method for solving optimization problems. Each method should be evaluated in the light of the problem solved.

In many cases it may be advantageous to use a combination of two methods - the first to obtain an approximate solution that would be used as starting functions for the next method. Summarized treatments of numerical methods, are given in /69 - 76/ and /11, 14, 48, 53, 58/; see also /77 - 79/ and for theoretical background /80/.
Let us make some final remarkes. We tacidly assumed in general as well for the analytical as for the numerical methods that a unique optimal control is existing. For example, in the special case of a singular solution the methods of second variations and quasilinearization cannot find singular trajectories; however, the gradient approach should converge to a singular solution if one exists. We did not pay any attention to problems in connection with singular solutions, or optimal control with state variable constraint.

3.6. Differential Games

The natural extension of the theory of (discrete) games, which we mentioned on page 8, to the dynamic case yields what is known as differential games. Mathematically it is a generalization of the conventional optimal control theory, at least in the most commonly treated "zero-sum" differential game in which there are two playes one of whom is trying to minimize a certain cost functional, the other to maximize it. We therefore have two types of control variables, $\underline{u}(t)$ and $\underline{v}(t)$. Suppose the composite dynamical system is given by the (time-independent) vector differential equation

$$\dot{\underline{x}} = \underline{f}\left[\underline{x}(t), \underline{u}(t), \underline{v}(t)\right], \quad \underline{x}(t_o) = \underline{x}_o \qquad (3.42)$$

where $\underline{x}(t)$ is the state vector. Furhter, let there be the usual types of boundary conditions and the constraints $\underline{u} \in \Omega$, $\underline{v} \in \psi$ on the problem.
Consider the cost functional to be of the form

$$J(\underline{u},\underline{v}) = \int_{t_o}^{t_f} F\left[\underline{x}(t), \underline{u}(t), \underline{v}(t)\right] dt \quad . \qquad (3.43)$$

The objective of the differential game is to find the optimal control functions $\underline{u}^*(t)$ and $\underline{v}^*(t)$ in $t_o \leq t \leq t_f$ - if they exist - such that on one hand the boundary conditions are satisfied and on the other hand the cost functional takes the form

$$J(\underline{u}^*, \underline{v}^*) = \min_{\underline{u} \in \Omega} \left\{ \max_{\underline{v} \in \psi} \int_{t_o}^{t_f} F\left[\underline{x}(t), \underline{u}(t), \underline{v}(t)\right] dt \right\}, \qquad (3.44)$$

e.g., the cost functional is "minimaximized". In other words, the player A who controls $\underline{u}(t)$ wishes to minimize the maximum possible value that the cost functional can attain due to the influence of the player B who controls $\underline{v}(t)$. The "saddle-point" condition for the game is

$$J(\underline{u}^*,\underline{v}) \leq J(\underline{u}^*,\underline{v}^*) \leq J(\underline{u},\underline{v}^*) \ . \qquad (3.45)$$

Under certain assumptions necessary conditions of the minimax problem posed above can be found by an extension of the maximum principle. By taking the usual Hamiltonian function

$$H\left[\underline{x}(t),\underline{p}(t),\underline{u}(t),\underline{v}(t)\right] = F\left[\underline{x}(t),\underline{u}(t),\underline{v}(t)\right] + \underline{p}^T(t)\,\underline{f}\left[\underline{x}(t),\underline{u}(t),\underline{v}(t)\right] \qquad (3.46)$$

and under the assumption that unique $\underline{u}(t)$ and $\underline{v}(t)$

exists such that $\max_{\underline{u}} \min_{\underline{v}} H = \min_{\underline{v}} \max_{\underline{u}} H$ we take

$$H^*\left[\underline{x}(t),\underline{p}(t)\right] = \min_{\underline{u}\in\Omega} \max_{\underline{v}\in\Psi} H\left[\underline{x}(t),\underline{p}(t),\underline{u}(t),\underline{v}(t)\right] \ . \qquad (3.47)$$

However, these results hinge upon the assumption that the differential game has a saddle-point, e.g., the minimax-value of the cost function (3.44) is equal to its maximum-value (separability assumptions); heuristically, it means that it makes no difference whether the player A or B plays first. The conditions for the above statement to hold are nontrivial, c.f. /11, 48, 81 - 83/.

As well the discrete as the differential games and in general certainly the optimization theory as a whole are important methods for adaptive and learning control systems, a class of problems which is beyond the scope of this paper /84 - 86 /.

3.7. Distributed Parameter Systems

A great many problems which occur in practice are characterized by distributed parameter systems. The dynamic behaviour of these systems is governed by sets of partial differential equations, integral equations, or integrodifferential equations, and sometimes by more general functional equations. In these systems we must characterize state variables not only as a function of time but in addition as a function of one or more spatial variables. It is by nature far more difficult to develop theories and techniques for optimal control of distributed parameter systems. But recently quite some efforts have been made in this field.

Most of the work in optimal control of distributed parameter systems is connected with the methods of the calculus of variations. For partial differential equations formulations, the necessary conditions take, analogous to lumped parameter systems, the form of canonical (Euler-Lagrange) equations together with a maximum principle and transversality conditions. The canonical equations are partial differential equations and the transversality conditions are boundary conditions for these equations. Whereas in the case of integral equation problems, the necessary condition typically takes the form of a linear or nonlinear integral equation which the optimum control must satisfy.

The following simplified version of the maximum principle for distributed parameter systems shows the analogy to that of lumped parameter systems /53/. Suppose the dynamical system is given by the partial vector differential equation

$$\frac{\partial \underline{x}(t,\underline{y})}{\partial t} = \underline{f}\left[\underline{y}, \underline{x}(t,\underline{y}), \frac{\partial^K \underline{x}(t,\underline{y})}{\partial \underline{y}^K}, \underline{u}(t,\underline{y})\right] \qquad (3.48)$$

with the initial and boundary conditions for the terms

$$\underline{x}(t_o,\underline{y}) = \underline{x}_o(\underline{y}) \qquad (3.49)$$

and

$$\frac{\partial^{K-1}\underline{x}(t,\underline{y})}{\partial \underline{y}^{k-1}}, \quad \frac{\partial^{K-2}\underline{x}(t,\underline{y})}{\partial \underline{y}^{K-2}}, \ldots, \quad \frac{\partial \underline{x}(t,\underline{y})}{\partial \underline{y}}, \underline{x}(t,\underline{y}) \qquad (3.50)$$

specified at t_o and the boundary $\partial\Omega$, where $\underline{y}$ is a m-dimensional spatial coordinate vector, $\underline{x}(t,\underline{y}) \in \Omega$ a n-dimensional state vector function, $\underline{u}(t,\underline{y})$ a r-dimensional distributed control vector, $\underline{f}$ a vector-valued function; the simple notation

$\frac{\partial^K \underline{x}(t,y)}{\partial \underline{y}^K}$ takes care of all possible existing partial

derivatives. We desire to find a control vector $\underline{u}(t,\underline{y})$ from among the admissible control region which minimizes the cost functions for fixed t_o and t_f

$$J = \int_{t_o}^{t_f} \int_\Omega F\left[\underline{y}, \underline{x}(t,\underline{y}), \frac{\partial^K \underline{x}(t,\underline{y})}{\partial \underline{y}^K}, \underline{u}(t,\underline{y})\right] d\,\Omega\, dt. \qquad (3.51)$$

If a solution exists, we obtain the two-point boundary value partial differential equation whose solution minimizes the cost function in an analogous manner as for lumped parameter systems by defining the Hamiltonian.

$$H = F + \underline{p}^T(t,\underline{y})\ \underline{f} \tag{3.52}$$

The canonical partial differential equations are obtained from

$$\frac{\partial \underline{x}}{\partial t} = \frac{\partial H}{\partial \underline{p}} = \underline{f} \tag{3.53}$$

$$\frac{\partial \underline{p}}{\partial t} = - \frac{\partial H}{\partial \underline{x}} - (-1)^K \frac{\partial^K}{\partial \underline{y}^K} \frac{\partial H}{\partial \left[\frac{\partial^K \underline{x}}{\partial \underline{y}^K}\right]}, \tag{3.54}$$

where the Hamiltonian is minimized with respect to choice of $\underline{u}$ such that

$$H^* \leq H; \tag{3.55}$$

the asterisk in eq. (3.55) again denotes that the optimal admissible control vector $\underline{u}^*(t,\underline{y})$ is used.

The knowledge of the maximum principle as we sketched it above may help an engineer to get a better intuitive insight into the structure of the optimal system. Knowledge or even a feeling of what can and what cannot be optimal is for the design of an optimal system of great importance. But, in general, for a real solution of an optimal problem the system must be reduced to a set of ordinary differential or difference equations.
We can for example discretizise this typ of problem in two principle ways. One way consists of obtaining the two-point boundary value partial differential equations and discretizing them. The other involves discretizing the spatial and/or temporal coordinates of the distributed parameter system itself. Certainly an advantage to the latter approach is that the physical system itself is approximated before any optimization is carried out. The optimization is then done on the approximate model, using, for example, the maximum principle for lumped parameter systems, which is considerably simpler to apply, than that of distributed parameter systems.
The last method is commonly used. From the few important ways in which this approximation may be carried out we like to name besides the spatial-, the time-, and the space-time discretization the eigenfunction expansion or harmonic truncation and the transfer function approximation.

We consider first the eigenfunction expansion. If it is assumed that the solution of a partial differential equation or an integral equation can be expressed as an infinite series of the form $x(t,\underline{y}) = \sum_i q_i(t) r_i(\underline{y})$ then it is possible in many cases to derive a set of

ordinary differential equations for the
$q_i(t)$, $i = 1,2,\ldots,\infty$. In many linear systems such a representation is valid and the $r_i(\underline{y})$ are the eigenfunction of the linear system. Of course, an infinite set of ordinary differential equations can, in general, not be solved either, but this set can be truncated at some point.
The transfer function method is more or less restricted to linear time-invariant distributed systems. In such distributed parameter systems the conventional transfer function representation consists of irrational functions, but they sometimes admit rational approximations. As it is with lumped systems, the computing techniques we presented in section 3.5 are still valid for discretized versions of distributed control problems. Extension of these techniques to distributed systems are considerably more complicated.

Recently the greater part of theoretical work in this field has centered on function space methods. In this case the problems are formulated in a suitable function space, e.g., Hilbert or Banach space. The advantage is, that one obtains very general results, which can be applied to a great class of problems.
Important questions like feedback solutions, identification problems, systems sensitivity which have received a great deal of attention for lumped systems have found also some interest. For a summarized treatment of optimal control of distributed-parameter systems /12,53,87,99/.

3.8. Stochastic Optimal Control

We have seen that the optimal control of a dynamic system requires a knowledge of the state of the system. In practice, however, the individual state variables cannot be determined exactly due to random errors of the measurement. The system itself may also be subjected to random errors. Thus, we are faced with the problem of making good estimates.

If the ideal system (perfect measurements and no random disturbances) is given, and if we have some knowledge of the degree of uncertainty in the measurements and of the degree of intensity of the random disturbances to the system, then, on the basis of all the measurements up to the present time, we can determine the most likely value of the state variables. The process of determining these most likely values is called smoothing, filtering or prediction, depending on whether we are finding past, present, or future values of state variables, respectively.

A well known class is the linear least-squares estimation and control based on the work of Wiener. In his work on steady-state filter he approached the problem by Fourier analysis and developed an integral equation(the Wiener-Hopf equation) for determining the impuls response matrix for the filter. A particular useful solution to the stationary and also nonstationary linear estimation problem was given by /90,91/, known as Kalman-Bucy filter theory. This work is based on optimal control theory such that the expected value (taken over the ensemble of all experiments) of the cost function is a minimum. Certainly the two methods in the frequency domain (Wiener) and in the time domain (Kalman) yield identical results for stationary problems.
A typical problem in stochastic optimal control is the synthesis of feedback controllers that are optimum in the ensemble average sense, in the presence of random disturbances and uncertainty in measurements and initial conditions. In most of the work linear systems with gaussian noise are considered. As an example, let us sketch a simple stochastic control problem that is designing an optimal controller for a linear system disturbed by gaussian white noise, where the performance index is quadratic, the initial conditions are random, but perfect knowledge of the state of the system is available /48/.

The system to be controlled is described by the linear matrix differential equation

$$\underline{\dot{x}}(t) = \underline{A}(t)\underline{x}(t) + \underline{B}(t)\,\underline{u}(t) + \underline{w}(t), \tag{3.56}$$

where $\underline{x}(t)$ is an n-dimensional state vector,

$\underline{u}(t)$ an r-dimensional control vector, and

$\underline{w}(t)$ a process noise vector with n-components.

We assume also

$$E\left[\underline{w}(t)\right] = 0, \text{ but } E\left[\underline{w}(t)\,\underline{w}^T(\tau)\right] = \underline{G}(t)\,\delta(t-\tau),$$

$$E\left[\underline{x}(t_o)\right] = 0, \text{ but } E\left[\underline{x}(t_o)\,\underline{x}^T(t_o)\right] = \underline{X}_o\,.$$

We wish to minimize the performance index, which we assume to be the ensemble average of a quadratic form like the one in eq. (3.6)

$$J = E\left\{\frac{1}{2}\,\underline{x}(t_f)\,\underline{S}\,\underline{x}(t_f) + \frac{1}{2}\int_{t_o}^{t_f} (\underline{x}^T\underline{Q}(t)\underline{x} + \underline{u}^T\underline{R}(t)\underline{u})dt\right\}, \tag{3.57}$$

where (in the absence of constraints on the structure of the controller) $\underline{S}$ and $\underline{Q}(t)$ are positive semidefinite

matrices and $\underline{R}(t)$ is a positive definite matrix.
Because $\underline{w}(t)$ was assumed to be a white noise, it is impossible to predict $\underline{w}(t)$ for $t > \tau$, even with perfect knowledge of the state for $\tau < t$. Then the optimal controller is identical to the deterministic controller (3.30) under the conditions (3.31) and (3.32).
It is often of interest to determine how the controller system will behave on the average. Therefore, we substitute eq. (3.30) into eq. (3.56) which yields

$$\dot{\underline{x}} = \left[\underline{A}(t) - \underline{B}(t)\, \underline{K}(t) \right] \underline{x}(t) + \underline{w}(t) \ . \qquad (3.58)$$

Let the mean sqare value of $\underline{x}(t)$ be

$$\underline{X}(t) = E \left[\underline{x}(t)\, \underline{x}^T(t) \right] , \qquad (3.59)$$

then one can show that the linear equation

$$\dot{\underline{X}} = \left[(\underline{A}(t) - \underline{B}(t)\, \underline{K}(t)\right] \underline{X} + \underline{X}\left[\underline{A}(t) - \underline{B}(t)\, \underline{K}(t) \right]^T + \underline{G}(t)$$

with the initial condition (3.60)

$$\underline{X}(t_o) = \underline{X}_o \qquad (3.61)$$

allows us to predict the mean square histories of the state variables and their cross-correlations.
In the statistically stationary case, that is the matrices $\underline{A}$, $\underline{B}$, $\underline{G}$, $\underline{Q}$ and $\underline{R}$ are constant, the mean-square value of the state $\underline{X}$, becomes constant and can be found - by setting in eq.(3.60) $\dot{\underline{X}} = \underline{O}$ - from the linear algebraic equation

$$(\underline{A} - \underline{B}\, \underline{K})\, \underline{X} + \underline{X}\, (\underline{A} - \underline{B}\, \underline{K})^T + \underline{G} = \underline{O} \qquad (3.62)$$

the mean-square value of the control also becomes constant. As we can see the optimal feedback-control algorithm presented in the preceding sections for deterministic linear systems also applies to optimal control of stochastic linear systems provided that we insert an optimal filter in the feedback loop. This is also true for discrete systems. However, stochastic problems involving partial differential equations have not had a great deal of attention.

For constraint optimal control problem one also forms averages in order to convert the original stochastic control problem into a deterministic problem. One uses then one of the proposed methods (maximum principle, dynamic programming) for its solution. For summarized representations see /13,48,53,92,93/.

3.9. Concluding Remarks

In this paper we did not at all attempt to treat the theory of optimal control as general as it is possible. It was more or less our purpose to point out some fundamental properties of optimal control and give a rough idea of different optimization principles. Almost no attention was given to the different for practical purpose important formulations such as time-optimal control, energy-optimal control or whether it is a terminal control problem or not (problems of Lagrange, Mayer, and Bolza) etc. We also did not consider any classification in regard with the application of optimal control.

As we said before, it is very difficult in practice to find the right performance index and a number of compromise steps need to be taken; thus we arrive at a suboptimal design. We desire to evolve a suboptimal design, that is nearly optimal. It is therefore, important how sensitive the design is to parameter and environmental changes. There are always in real systems additional disturbances entering the optimal system which were not considered in the optimization process. In most cases one makes great efforts in order to get an optimal closed-loop system, because it will take into account that a disturbance has occured.

We should not close this paper without making a few remarks in respect to multivariable systems. By formulating all our problems in form of vector equations we did not make any distinction between the optimization of multivariable and single variable systems. At first glance one might think that the difference between the two types of systems is only one of degree, that is with some extra calculation one could extend single variable methods to multivariable systems. Unfortunately this is not true, multivariable problems can have an entirely differnt form or structure than a single variable one. One sometimes refers to the difficulties which are generated by these differences as "the curse of dimensionality" /46/. Because of the more complex structure of general multivariable systems it is far more difficult to find even a good suboptimal solution by engineering intuition.

A class of systems which find an increased interest are the multilevel control systems. The optimal control of these large-scale, complex systems which consists of many dynamic and interacting subsystems, represents a difficult problem for the control engineer. This situation is partly due to the fact that the existing optimization theories usually deal with isolated, simple systems and the available controllers or computers are frequently designed or programmed for single processes.

A basic principle of the multilevel control of complex dynamic systems is the decomposition techniques, which we already mentioned in the linear and convex programming methods. In the class of problems, in which a number of independent, optimally controlled processes or operations is being aggregated in order to complete a whole hyperprocess or project, a higher level controller is being used in order to ensure the optimum coordination of all the operations that constitute the project /31, 94/.

References and Bibliography.

/ 1/ Akhiczer, N.I.: The Calculus of Variations. Blaisdell Publ.Co, New York, 1962.

/ 2/ Bliss, G.A.: Lectures on the Calculus of Variations. The University of Chicago Press, Chicago, 1963.

/ 3/ Bolza, O.: Vorlesung über Variationsrechnung. Chelsea Publ.Co., New York, 1909.

/ 4/ Courant, R. and Hilbert, D.: Methods of Mathematical Physics. Vol. 2. Interscience Publishers, New York 1962. German ed.: Methoden der mathematischen Physik. Springer Verlag, Berlin, 1968.

/ 5/ Frank, W.: Mathematische Grundlagen der Optimierung. R. Oldenbourg Verlag, München, 1969.

/ 6/ Funk, P.: Variationsrechnung und ihre Anwendung in Physik und Technik. Springer Verlag, Berlin, 1962.

/ 7/ Grüss, G.: Variationsrechnung. Quelle und Meyer, Heidelberg, 1955.

/ 8/ Hestenes, H.R.: Calculus of Variations and Optimal Control. John Wiley a. Sons, Inc., New York, 1966.

/ 9/ Lee, E.B. and Markus, L.: Foundations of Optimal Control Theory. John Wiley a. Sons. Inc., New York, 1967.

/10/ Bauer H. and Neumann K.: Berechnung optimaler Steuerungen - Maximumprinzip und dynamische Optimierung. Lecture Notes in Operations Research and Mathematical Systems No. 17. Springer-Verlag, Berlin, 1969.

/11/ Athans, M.: The Status of Optimal Control Theory and Applications for Deterministic Systems. IEEE-Trans. on Automatic Control, Vol.AC-11 (1966), pp.580-596.

/12/ Robinson, A.C.: A Survey of Optimal Control of Distributed-Parameter Systems. Automatica., Vol 7, (1971), pp.371-388.

/13/ Kushner, H.J.: On the Status of Optimal Control and Stability for Stochastic Systems. 1966 IEEE International Convention Record, pt.6, pp.143-151.

/14/ Paiewonsky, B.: Optimal Control: A Review of Theory and Practice. AIAA-Journal, Vol. 3 (1965), pp. 1985 - 2006.

/15/ Dantzig, G.B.: Linear Programming and Extensions. Princeton Univ.Press, Princeton N.J.1963. German ed.: Lineare Programmierung und Erweiterungen. Springer-Verlag, Berlin,1966.

/16/ Charnes, A., Cooper, W.W. and Henderson, A.: An Introduction to Linear Programming. John Wiley and Sons, Inc., New York, 1953.

/17/ Hadley, G.: Linear Programming. Addison-Wesley Publ. Co., Reading, Mass., 1962.

/18/ Künzi, H.P.: Tzschach, H.G. and Zehnder, C.A.: Numerische Methoden der mathematischen Optimierung. B.G. Taubner Verlag, Stuttgart, 1966.

/19/ Dantzig, G.B. and Wolfe, P.: Decomposition Principle for Linear Programming. Operations Research, Vol. 8 (1960) pp. 101-111.

/20/ Künzi, H.P. and Tan, S.T.: Lineare Optimierung großer Systeme. Lecture Notes in Mathematics. No.27. Springer-Verlag, Berlin, 1966.

/21/ Dantzig, G.B.: Linear Programming under Uncertainty. Management Science, Vol.1 (1955), pp. 197-206.

/22/ Madansky, A.: Linear Programming under Uncertainty. In Wolfe, Ph. and Graves. R.L. (eds.): Recent Advances in Mathematical Programming. McGraw-Hill Book Co., New York, 1963.

/23/ Faber, M.M.: Stochastisches Programmieren. Physica-Verlag, Würzburg, 1970.

/24/ Kuhn, H.W. and Tucker, A.W.: Nonlinear Programming. Proceeding of the Second Berkeley Symposium on Mathematical Statistics and Probability. Univ.of California Press, Berkeley, Calif.,1951.

/25/ Künzi, H.P. and Krelle, W.: Nichtlineare Programmierung. Springer-Verlag, Berlin, 1962.

/26/ Collatz, L. and Wetterling, W.: Optimierungsaufgaben. Springer-Verlag, Berlin, 1966.

/27/ Vajda, S.: Mathematical Programming. Addison-Wesley Publ. Co., Reading, Mass., 1961.

/28/ Wolfe, Ph.: The Simplex Method for Quadratic Programming. Econometrica, Vol. 27 (1959), pp, 382-398.

/29/ Wolfe, P.: Methods of Nonlinear Programming. In Graves, R.L. and Wolfe, P. (eds.): Recent Advances in Mathematical Programming. McGraw-Hill Book Co., New York, 1963.

/30/ Varaiya, P.P.: Decomposition of Large-scale Systems.. In Zadeh, L.A. and Polak, E.(eds.): System Theory. McGraw-Hill Book Co., New York, 1969.

/31/ Baumann, E.J.: Multilevel Optimization Techniques with Application to Trajectory Decomposition. In Leondes, C.T. (ed.): Advances in Control Systems, Vol. 6. Academic Press, New York, 1968.

/32/ Pierre, D.A.: Optimization Theory with Applications. John Wiley and Sons, Inc., New York, 1969.

/33/ Wilde, D.J.: Optimum Seeking Methods. Prentice-Hall, Inc., Englewood Cliffs, N.J., 1964.

/34/ Wilde, D.J. and Beightler, C.S.: Foundations of Optimizations. Prentice-Hall, Inc., Englewood Cliffs, N.J., 1967.

/35/ Fiacco, A.V. and McCormick,G.P.: Nonlinear Programming: Sequential Unconstraint Minimization Techniques. John Wiley and Sons, Inc., New York, 1968.

/36/ Draper, C.S. and Li, Y.T.: Principles of Optimalizing Control Systems and an Application to the Internal Combustion Engine. American Society of Mechanical Engineers, New York, 1951.

/37/ Feldbaum, A.A.: Rechengeräte in automatischen Systemen. R. Oldenbourg Verlag,München, 1962.

/38/ Mishkin, E. and Braun, Jr.L. (eds.): Adaptive Control Systems. McGraw-Hill Book Co., New York, 1961.

/39/ Eveleigh, V.W.: Adaptive Control and Optimization Techniques.McGraw-Hill Book Co., New York, 1967.

/40/ Lee, T.H., Adams,G.E. and Gaines,W.M.: Computer Process Control: Modeling and Optimization. John Wiley and Sons, Inc., New York, 1968.

/41/ Neumann, J.von, and Morgenstern, O.: Theory of Games and Economic Behaviour, Princeton Univ. Press, Princeton, N.J., 1953. German ed.: Spieltheorie und wirtschaftliches Verhalten. Physica-Verlag, Würzburg, 1961.

/42/ Burger, E.: Einführung in die Theorie der Spiele. W. de Gruyter-Verlag, Berlin, 1959.

/43/ McKinsey,J.: An Introduction to the Theory of Games. McGraw-Hill Book Co., New York, 1952.

/44/ Blaquiere, A. and Leitmann, G.: Multiple-Stage Quantitative Games. In Moiseev, N.N. (ed.): Colloquium on Methods of Optimization. Lecture Notes in Mathematics No.112, Springer-Verlag, Berlin, 1970.

/45/ Pontryagin, L.S., Boltyanskii, V.G., Gamkrelidze, R.V. and Mishchenko, E.F.: Mathematical Theory of Optimal Processes. John Wiley a.Sons, Inc., New York, 1962.
German ed.: Mathematische Theorie optimaler Prozesse. R. Oldenbourg-Verlag, München,1964.

/46/ Bellman, R.: Dynamic Programming. Princeton University Press, Princeton, N.J., 1957.

/47/ Athans, M. and Falb, P.L.: Optimal Control. McGraw-Hill Book Co., New York, 1966.

/48/ Bryson,Jr.,A.E. and Ho,Y.C.: Applied Optimal Control. Blaisdell Publ. Co., Waltham,Mass., 1969.

/49/ Feldbaum, A.A.: Optimal Control Systems. Academic Press, New York, 1965.

/50/ Lapidus, L. and Luus, R.: Optimal Control of Engineering Processes. Blaisdell Publ.Co., Waltham, Mass., 1967.

/51/ Leitmann, G. (ed.): Topics in Optimization. Academic Press, New York, 1967.

/52/ Naslin, P.: Essentials of Optimal Control. Boston Technical Publishers, Boston, 1969.

/53/ Sage, A.P.: Optimum Systems Control. Prentice-Hall, Inc., Englewood Cliffs, N.J., 1968.

/54/ Fan, L.T.: The Continuous Maximum Principle. John Wiley a.Sons, Inc., New York, 1966.

/55/ Bellman, R. and Kalaba, R.: Dynamic Programming and Modern Control Theory. Academic Press, New York, 1965.

/56/ Dreyfus, S.E.: Dynamic Programming and the Calculus of Variations. Academic Press, New York,1965.

/57/ Elgerd,O.I.: Control Systems Theory. McGraw-Hill Book Co., New York, 1967.

/58/ Hsu,J.C. and Meyer,A.V.: Modern Control Principles and Applications. McGraw-Hill Book Co., New York, 1968 .

/59/ Takahashi, Y., Rabins, M.J. and Auslander,D.M.: Control and Dynamic Systems.Addison-Wesley Publ.Co., Reading, Mass., 1970.

/60/ Tou,J.T.: Modern Control Theory.McGraw-Hill Book Co., New York, 1964.

/61/ Halkin,H.: Optimal Control for Systems Described by Difference Equations. In Leondes,C.T. (ed.): Advances in Control Systems. Academic Press, New York, 1964.

/62/ Holtzman, J.M., Convexity and the Maximum Principle for Discrete Systems. IEEE Trans. on Automatic Control.Vol.AC-11 (1966),pp.30-35.

/63/ Anderson, B.D.O. and Moore, J.B.: Linear Optimal Control. Prentice-Hall, Inc., Englewood Cliffs, N.J., 1971.

/64/ Kalman,R.E.: The Theory of Optimal Control and the Calculus of Variations. In Bellman,R. (ed.): Mathematical Optimization Techniques. University Press of California, Berkeley, Calif., 1963.

/65/ Luenberger, D.G.: Observers for Multivariable Systems. IEEE Trans.on Automatic Control. Vol.AC-11.(1966),pp.190-197.

/66/ Roberts, S.M.: Dynamic Programming in Chemical Engineering and Process Control, Academic Press, New York, 1964.

/67/ Larson, R.E.: State Increment Dynamic Programming American Elsevier Publ.Co., Inc., New York,1968.

/68/ Bellman, R., Kalba, R. and Kotkin, B.: Polynomical Approximation - A New Computational Technique in Dynamic Programming - I, Allocation Processes. Mathematics of Computation.Vol.17(1963),pp.155-161.

/69/ Isaacs, D.: Algorithms for Sequential Optimization of Control Systems. In Leondes, C.T. (ed.): Advances in Control Systems, Vol. 4. Academic Press, New York, 1966.

/70/ Kenneth, P. and McGill,R.: Two-Point Boundary-Value-Problem Techniques. In Leondes, C.T. (ed.) Advances in Control Systems, Vol. 3. Academic Press, New York, 1966.

/71/ Koop, R.E. and Moyer, G.H.: Trajectory Optimization Techniques. In Leondes, C.T.(ed.): Advances in Control Systems, Vol. 4. Academic Press, New York, 1966.

/72/ Payne, J.A.: Computational Methods in Optimal Control Problems. In Leondes, C.T.(ed.): Advances in Control Systems, Vol.7. Academic Press, New York, 1969.

/73/ Scharmack, D.K.: An Initial Value Method for Trajecotory Optimization Problems. In Leondes, C.T.:(ed.): Advances in Control Systems, Vol. 5. Academic Press, New York 1967.

/74/ Bellman, R.E. and Kalaba, R.E.: Quasilinearization and Nonlinear Boundary-Value Problems. Elsevier Publ. Co., Inc., New York 1965.

/75/ Bailey, P.B., Shampine, L.F. and Waltman, P.E.: Nonlinear Two Point Boundary Value Problems. Academic Press, New York, 1968.

/76/ Plant, J.B.: Some iterative Solutions in Optimal Control. Research Monograph No.44, The M.I.T. Press, Cambridge, Mass., 1968.

/77/ Balakrishnan,A.V. and Neustadt, L.W. (eds.): Computing Methods in Optimization Problems. Academic Press, New York, 1964.

/78/ Beckmann, M. and Künzi, H.P. (eds.): Computing Methods in Optimization Problems. Lecture Notes in Operations Research and Mathematical Economics, No. 14. Springer-Verlag Berlin 1969.

/79/ Mufti, I.H.: Computational Methods in Optimal Control Problems. Lecture Notes in Operation Research and Mathematical Systems, No.17, Springer Verlag, Berlin 1970.

/80/ Kantarovich, L.V. and Akilov, G.P.: Functional Analysis in Normed Spaces. The Macmillan Co., New York, 1964.

/81/ Isaacs, R.: Differential Games. John Wiley a.Sons, Inc., New York, 1965.

/82/ Ho, Y.C.: Differential Games and Optimal Control Theory. Proc.NEC, Vol.21(1965), pp.613-615.

/83/ Berkovitz, L.D.: Variational Approach to Differential Games. In Advances in Game Theory-Princeton Univ. Press, N.J.,Ann. Math. Studies, No.52 (1964), pp.127 - 173.

/84/ Eveleigh, V.W. Adaptive Control and Optimization Techniques. McGraw-Hill, Book Co., New York 1967.

/85/ Sworder, D. : Optimal Adaptive Control Systems, Academic Press, New York, 1966.

/86/ Tsypkin , Ya.Z.: Self-Learning - What is it? IEEE Trans.on Automatic Control AC-13 (1968) pp. 608 - 612.

/87/ Wang, P.K.C.: Control of Distributed Parameter Systems. In Leondes, C.T.(ed.): Advances in Control Systems, Academic Press, New York, 1964.

/88/ Brogan, W.L.: Optimal Control Theory Applied to Systems Described by Partial Differential Equations. In Leondes, C.T.(ed.): Advances in in Control Systems, Academic Press, New York, 1968.

/89/ Butkovsky , A.G., Egorov, A.I. and Lurie, K.A.: Optimal Control of Distributed Systems (A Survey of Soviet Publications). SIAM Journal on Control, Vol.6 (1968), pp. 437 - 476.

/90 / Kalman, R.E.: A new Approach to Linear Filtering and Prediction Problems. J.Basic Eng. Vol.82 (1960), pp.35-45.

/91/ Kalman, R.E. and Bucy, R.S.: New Results in Linear Filtering and Prediction Theory. J. Basic Eng. Vol. 83 (1961), pp.95-108.

/92/ Johansen, D.E.: Optimal Control of Linear Stochastic Systems with Complexity Constraints. In Leondes, C.T. (ed.): Advances in Control Systems, Academic Press, New York, 1966.

/93/ Kushner, H.J.: Stochastic Stability and Control. Academic Press, New York, 1967.

/94/ Kulikowski, R.: Optimum Control of Multidimensional and Multilevel Systems. In Leondes, C.T.(ed.): Advances in Control Systems, Academic Press, 1966.

M.D. Mesarovic

MULTILEVEL SYSTEMS THEORY: STATE OF THE ART, 1971

MULTILEVEL SYSTEMS THEORY: STATE OF THE ART, 1971

M.D. Mesarovic
Director,
Systems Research Center
Case Western Reserve University
Cleveland, Ohio, U.S.A.

SUMMARY

Main developments in the multilevel systems theory are reviewed with the discussion oriented toward applications. Some case studies are presented to show typical problems on which the present theory can be applied. The paper concludes with the suggestion of the areas for further research and development.

INTRODUCTION

A multilevel control system is a system which contains a family of control units arranged in a hierarchical fashion; by this we mean, in the most general case, that there exists at least one control unit which does not have a direct access to the controlled process but rather controls some other control units. Examples of different multilevel control systems will be given in the subsequent section; more complete discussion of the multilevel notions can be found in /19/.

Multilevel systems are not a novelty either in engineering or in scientific studies; however multilevel systems theory is a new development. It was founded in early sixties by / 1/ and / 2/ where the framework was provided for further research effort; this was followed by a steady flow of contributions of which / 3/ - /16/ represent a selection. Finally, the book /19/ summarized most of the developments in a unified form and also presented many previously unpublished results.

We shall make no effort to survey the multilevel systems theory in all its details; for that we have neither space nor time. Rather, we shall present a selection of developments in several diverse areas of multilevel research to show what one can do today by using multilevel systems theory.

There has been much concern recently about the gap between the control theory and practice. I maintain that one of the reasons for the gap is that the theoretical research is not concerned, by and large, with the issues and problems of primary importance for practice. It is not the mathematical framework of the theoretical developments or the way the findings are reported which is responsible for the gap; rather, the relative irrelevance of the issues considered creates the gap. Some of the most important problems in modern control technology are in the area of large scale and complex systems. How to proceed in an orderly fashion in the design of the control system for an integrated automation of a plant which involves not only physical control but scheduling, operation control, planning and other economic objectives? These are the questions for which some insight is very much in need today and for which multilevel systems theory is intended. To satisfy the needs for practical applications much more research is needed in the future. Such developments would contribute significantly to close the above mentioned gap.

BASIC TRENDS

There are two basic areas of research in multilevel systems theory: First deals with the question of structuring a complex control system which has to operate under conditions of a high degree of uncertainty; Seond deals with the question of coordination, i.e.: how to control a set of interacting control subsystems so as to harmonize their operations, i.e., avoid conflict and advance the overall system's objective. The first area of structuring and design of complex control system is based on engineering intuition and experience as well as on the study of the real-life multilevel systems in various fields. The theory presently available is conceptual rather than mathematical. The second area - coordination theory - has reached the stage where the problems and issues are stated with sufficient precision for mathematical investigations. Indeed, many interesting and difficult results have been already achieved.

Design of Complex Control Systems

Starting point for the design of complex control systems is provided by the recognition of several distinct and basic types of multilevel (or hierarchical) structures. First, there is a hierarchy resulting from the need to describe a complex system under control from variety of different angles, and by using different degrees of aggregation of variables, i.e., describing the controlled system in various degree of details. For example, a complex industrial process (such as refinery, still-mill, electric power system, etc.) is usefully described for control purposes on the level of physical variables, on the level of information handling (scheduling, monitoring, etc.) and on the level of economics on which the variables and operation of subprocesses are described in terms of their economic values and contribution to the profit and cost considerations. See figure 1. This kind of levels are referred to as *strata* and the total description of the process is denoted as a *stratified description*. The second kind of hierarchies appear as part of a strategy to control a complex process by using techniques developed for the simpler types of problems. The strategy results in a vertical decomposition of the overall control problem into different levels of decision complexity, referred to as *decision layers*. For example, see figure 2 in the process control area it is useful to distinguish *direct control layer*, *process control* (or optimization) *layer*, *adaptation* (or up-dating) *layer* and *self-organization* (or structuring) *layer*. The third kind of hierarchies appears in a complex control system where there is a number of control units working in "parallel", (i.e., controlling the subprocesses which are interacting), and to promote the overall objective it is needed to have an additional unit with the task of coordinating the operation of the "parallel" units. This is referred to as *organizational hierarchy* (also *multilevel-multigoal hierarchy*, or multiechelon hierarchy). It should be noticed that in the development of a stratified system's model, or a multi-layer control system one is concerned only with what might be called vertical decomposition, i.e., relationship between subsystems on different levels in a hierarchy while the organizational hierarchy is based on the horizontal decomposition as well, i.e., the existence of more than one unit on a given level.

In the design of a complex control system one uses all three types of hierarchies at different stages of the design process and in different ways. For example, the design process would start with a stratified description of the system and the identification of decision layers while the implementation is based on an organizational hierarchy in order to use the available equipment and observe the cost and time constraints. Actually, the interplay between various hierarchical concept is much more intricate than this brief discussion could imply. For example, after the tasks for various decision layers have been formulated the control unit on a given layer can use a stratified description of the control system suitable for its own purpose; or a unit in an organizational hierarchy, (such as e.g. the process optimization computer) can develop decision-layers appropriate for its own task.

Characteristic properties of various hierarchical arrangements useful for the control system design has been recognized and presented in /19/ or / 7/. At present they add up to a design procedure which for most effective application has to rely rather heavily on the understanding of the particular process under consideration. As an illustration a practical application to the design of a traffic control system is presented in a subsequent section. The system discussed has been implemented and evaluated in actual operation.

Coordination Theory

For the sake of simplicity we shall present the basic coordination problem by using ordinary differential equations model of the controlled process. The problem formulation however is much more general and is valid, for example, for partial differential equations models, algorithmic description of the processes, etc.

Let there be given n controlled processes with the models described by

$$\frac{dy_i}{dt} = f_i(y_i, m_i, u_i, \alpha_i) \ , \ i = 1,\dots,n$$

where y_i is the output, m_i the control, u_i the interaction and α_i the intervention (or coordination) parameter; each of these variables is, in general, a vector of time functions.

Each of the subsystems is controlled according to a performance function

$$v_i = g_i(m_i, y_i, \beta_i)$$

where g_i is usually a functional, i.e., v_i is a real number or a set of numbers in case of a multivalued objective.

The total output of the entire control system will be denoted by

$y = (y_1, \ldots, y_n)$

and similarly for control and interaction

$m = (m_1, \ldots, m_n)$

$u = (u_1, \ldots, u_n)$

The model of the overall controlled process is then represented by

$y = f(m)$

Interaction vector, u_i, depends upon the controls of the remaining subsystems, which for the simplicity of notation will be denoted by

$u_i = K_i(m)$

For the evaluation of the control of the overall process there is a performance (evaluation) function

$v = g(y,m)$

The control situation is now the following: The process control variables $m_1, \ldots, m_n$ are selected by the respective first level control units on the basis of the local process model, say f_i, and in reference to the local objective function g_i. There is a second level unit concerned with the overall performance function, g, and the total process f. Since the process control variables $m_1, \ldots, m_n$ are selected by the first level units the second level controller can affect that choice, in order to advance the overall objective, only by a proper selection of the coordination vectors

$\alpha = (\alpha_1, \ldots, \alpha_n)$ and $\beta = (\beta_1, \ldots, \beta_n)$

This activity is referred to as coordination. To arrive at a strategy for coordination the following two questions have to be answered:

(i) How should the first level unit consider the interactions? Namely, since m_i is selected on the basis of f_i, the information about u_i has to be given to the respective control unit.

(ii) On what basis should the second level unit select the coordination variables α and β?

For the solution of the first question the following approaches are proposed:

(a) *Decoupling (coordination) mode:* Each of the first level units is requested to consider the interaction as an additional control variable so that in addition to selecting the control, denoted by $\tilde{m}_i$, it generates the information about the interaction desirable for its own local operation, denoted by $\tilde{u}_i$. Of course, only $\tilde{m}_i$ can be applied to the process and in general $\tilde{u}_i$ is different than the actual interaction, i.e.,

$\tilde{u}_i \neq K_i(\tilde{m})$

(b) *Prediction (coordination) mode:* The second level controller communicates to the first level units the (future) values of the interaction variables, denoted by, $\alpha_1,\ldots,\alpha_n$, so that the first level controllers can concentrate on the local concerns since the controlled process is effectively decoupled as far as the first level units are concerned.

(c) *Disturbance (coordination) mode:* The first level units are instructed to consider the interaction variables as disturbances, possibly constrained to within a given set; i.e., the first level units have to select local controls under assumption than any interaction, from within a given set, can actually happen in the future.

Regarding the answer to the second question, i.e., what ought to be the strategy for coordination there are two lines of developments: via the so-called coordination principle and the heuristic approach.

The coordination principles are introduced in reference to various coordination modes, in particular:

(α) For the decoupling mode the second level unit changes theintervention parameters, in this case β only, until the actual values of interactions, say $K_i(\tilde{m})$, become precisely the same as the interactions derived by the first level units, i.e., until the following equation is satisfied.

$\tilde{u}_i = K_i(\tilde{m})$

This is referred to as *interaction balance principle* (of coordination).

(β) For the prediction mode, the second level unit changes the value of the predicted interaction, say α_i, until it becomes the same as the one which actually takes place, i.e.,

$\alpha_i = K_i(\tilde{m})$

Notice that β does not necessarily effect this approach. It is referred to as *interaction prediction principle* (of coordination).

(γ) For the disturbance mode the second level unit changes the intervention parameters, β in this case, until actual interactions fall within the prescribed bounds. This is referred to the *interaction estimation principle* (of coordination).

In the use of coordination principles as a basis for coordination the issues of interest are in particular the following:

(1) For what kinds of systems a given coordination principle leads to successful coordination? By this we mean that there exists a value for the coordination variables (α,β) so that the control selected by the first level units satisfy the overall objective as evaluated by g.

(2) How to design practical procedures for the selection of the coordination variables (α,β)?

Both of these issues have been investigated in considerable details, especially the first one. The results are too numerous to review here. Suffices to say that all three coordination principles are applicable under rather broad conditions. As an illustration of the type of the results available, we have for the interaction balance principle the following:

Theorem: *Let each of the subprocesses be described by a vector differential equation*

$$\frac{dy_i}{dt} = f_i(m_i,u_i) \ , \ i=1,\ldots,n$$

where f_i is a nonlinear function, the local performance function is given by

$$g_{i\beta} = g_i(m_i,u_i) + \beta_i u_i - \sum_{j=1}^{n} \beta_j K_j(m)$$

and the overall performance function is the sum

$$g = g_1(m_1,u_o) + \ldots + g_n(m_n,u_n)$$

Let the overall objective be to minimize g while the objective of each of the first level units is to minimize the respective local performance, $g_{i\beta}$. Then the system is coordinable by the interaction balance principle if and only if there exists a coordination parameter, β, such that

$$\min_{m,u} \sum_{i=1}^{n} g_{i\beta}(m_i,u_i) = \min_{m} g(m) \qquad (A)$$

If certain differentiability conditions are satisfied the condition (A) from the preceding theorem is satisfied if the first level performance functions $g_{i\beta}$ are convex functions and the control sets $M_i \times U_i$ are convex sets. These are fairly general conditions and therefore the interaction balance principle provides a very general strategy for coordination.

Actually the interaction balance principle is applicable even for a broader class of systems. Detailed discussion of these results as well as the investigation of the applicability of other principles can be found in /19/.

It is of interest here to compare coordination procedures based on the three principles. In this respect the following statements can be made:

(α) Coordination by the interaction balance principle is applicable more broadly than for the prediction case.

(β) Coordination by prediction is easier for the first level units and can be also adopted for various heuristic approaches; e.g. the second level can solve the overall but simplified problem and specify the interactions on that basis and let the first level units finally select the process control variables by an improved model but on a local basis.

(iii) Coordination by interaction estimation is suitable for on-line application and by using various adaptive type techniques.

Regarding the issue (2), i.e., finding a practical search procedure for the coordination process, two approaches have been proposed: Iteration approach for the optimization case and the second level feedback approach for the on-line case in the presence of disturbances. The iteration approach has been studied in some depth and the evidence strongly indicates the superiority of multilevel approach. If the computer storage is the limiting factor the advantage of multilevel approach is obvious. Less obvious, but still true is the advantage obtained as a reduction of computer time. This advantage, of course, can not be proven with a mathematical rigour since it depends upon the specific structure of the problem involved. Results from a case study will be discussed later.

In addition to coordination principles, several heuristic procedures have been proposed for the coordination strategy. An excellent example is given in the paper by B. Cheneveaux /21/ presented at this meeting I will not go into details of this or other heuristic approaches. However it should be pointed out that most of them can be shown to be derivatives from the coordination principles, i.e., represent an approximation to a coordination strategy derived from the application of a given coordination principle.

APPLICATIONS

To indicate the range of applications of already existing theories we shall present one example from each of the areas of research discussed in the preceding section.

Design of a Freeway Control System

To speed the traffic from one side of a large city to another urban freeways are built which have a limited access via controlled ramps. The problem in operation as well as in deciding on the construction of new freeway systems is that they are in excessive use during short peak hours of heavy traffic and are used below capacity during the rest of time. There is obviously here a control problem: to control the access to the freeway so as to maximize the traffic flow during peak hours. This is done by appropriately changing the ramp signals. The measurement variables include: speed of the vehicles, gap between the consecutive vehicles, merging characteristics, etc. The control problem is a very complex one because of the unpredictable response of the "controlled system", need for a fast action and the effect of a large number of unpredictable events, such as the occurrence of an accident causing congestion.

Using multilevel systems approach discussed in the preceding section a computer control system was built and implemented /20/. The layer decomposition of the overall control function is shown in figure 4. Specific control function on each of the layers is indicated in the diagram. Each of the control layer is in itself a rather complicated control system. Discussion of these is outside of the scope of this presentation but for the sake of illustration the diagram of the respective control system on each of the layers is given in figure 5, 6 and 7.

An additional practical advantage of the multilevel approach to complex control systems ought to be mentioned: Although the initial design includes the total control system the implementation can proceed in stages; i.e., starting from the lowest level the additional levels can be added when the technical, operational or economic conditions warrant.

Optimal Scheduling of Hydro-Thermal Power Systems

Scheduling of the power generation in a hydro-thermal power system is a very complex and large optimization problem further complicated by a large number of constraints for physical and security reasons and the need to take into account dynamics of the hydro subsystem. To alleviate the computational problem a method has been developed to apply multilevel approach to the solution of the scheduling problem /17/. Method essentially utilizes the decomposition of the system in two subsystems. S_1 - which contains the hydro part and S_2 - which contains the electric network and the thermo power stations. For each of the subsystems there is a separate decision unit optimizing on alocal basis. To coordinate local solutions decoupling mode of coordination is used with the hydro power as the interaction variable, incremental cost of hydro power as the coordination parameter and the interaction balance principle as the basis for coordination strategy. The solution procedure is then essentially the following: The coordinator specifies the initial incremental cost λ and the first level units derive the hydro power on the local basis, i.e., S_1 determines the power which it is proposed to deliver, P_{HG}, while S_2 specifies the hydro power it wants to buy, P_{HD}. If there is a difference, $P_{HG} - P_{HD} \neq 0$, the coordinator change the cost λ until the difference is eliminated.

In addition to obvious advantage of working on two subproblems of smaller dimensions there is the advantage due to the fact that the dynamics of the system is separated in S_1. In various actual applications further decomposition of the subsystems might be desirable.

There has been several other applications of coordination principles.

Applications of the coordination principles in a number of other areas have been reported. E.g. in an application to water quality control a 40% saving in computer time to determine the optimum control and regulation strategy has been reported /12/.

CONCLUSION

The range of research opportunities in the multilevel control area is very broad and we shall indicate here only those which seem to be of greatest immediate need and also offering good prospect for arriving at satisfactory solution.

In the area of theoretical research the problems of interest include:

(i) Development of practical algorithms for the selection of coordination parameters.

(ii) Extension of the coordination principles to distributed parameters systems (in particular in reference to /18/).

(iii) Development of suboptimal, heuristic procedures for coordination.

(iv) Introduction of the stochastic concepts in the coordination theory.

(v) Development of adaptive coordination procedures.

(vi) Extension of the theory to systems with more than two levels.

In the area of application the greatest need is still in the problem formulation area. Namely, there exists a range of concepts which can be used in structuring large scale integrated automation systems and their applicability has not been fully exploited.

REFERENCES

/1/ Mesarovic, M.D., A General Systems Approach to Organization Theory. Systems Research Center Report SRC 2-A-62-2, 1962.

/2/ Mesarovic, M.D., A Conceptual Framework for the Studies of Multilevel multigoal Systems. Systems Research Center Report SRC 101-A-66-43, 1966.

/3/ M.D. Mesarovic, J. Sanders, C. Sprague, An Axiomatic Approach to Organizations from a General Systems Viewpoint in "New Perspectives in Organization Research", Cooper, Leavitt and Shelly, editors, J. Wiley. New York, 1962.

/4/ Lasdon, L. and Schoeffler, J.D., A Multilevel Technique for Optimization. Proc. JACC, Troy, New York, 1965.

/5/ Takahara, Y. and Mesarovic, M.D., Coordinability of Dynamic Systems. IEEE Transactions on Automatic Control, Vol. AC-14, No. 6, 1969.

/6/ Mesarovic, M.D., Macko, D., Takahara, Y., Two Coordination Principles and Their Application in Large-Scale Systems Control. IV IFAC Congr., Warsaw, Poland, 1969.

/7/ Lefkowitz, I., Multilevel Approach Applied to Control System Design. Trans. ASME 88D, 2, 1966.

/8/ Kulikowski, R., Optimization of Large-Scale Systems, IV Congress IFAC, Warszawa, 1969.

/9/ Straszak, A., Optimal and Suboptimal Multivariable Control Systems With Controller Cost Constraint. Proc. III IFAC Congr., London, 1966.

/10/ Brosilow, C.B., Lasdon, S., Pearson, J.D., Feasible Optimization Methods for Interconnected Systems. Proc. JACC, Troy, New York, 1965.

/11/ Wismer, D.A., Optimal Control of Distributed Parameter Systems Using Multilevel Techniques, Ph.D. Thesis, UCLA, 1966.

/12/ Haimes, Y.Y., Modeling and Control of the Pollution of Water Resources Systems via Multilevel Approach, Water Resources Bulletin, 1971.

/13/ Mesarovic, M.D., Multilevel Systems and Concepts in Process Control, IEEE Proceedings, 1970.

/14/ Nicholson, H., Hierarchal Control of a Multimachine Power System Model, IEEE Winter Power Meeting, 1968.

/15/ Findeisen, W., Multilevel Systems of Control, Automatica and Telemehanica, 1970.

/16/ Aronovitch, V.V., Gavrilov, P., Romm, R.F., Hierarchical Algorithms of Optimal Control of Chemical-Technological Complexes, Automatica and Telemechanica, 1969.

/17/ Bonaert, A.P., El-Abiad, H., Koivo, A.J., On the Optimal Scheduling of Hydro-Thermal Power Systems, Report TR-EE 70-21, Purdue University, 1971.

/18/ Lions, J.L., Temam, R., Eclatement et Decentralisation en Calcul des Variations, to be published.

/19/ Mesarovic, M.D., Macko, D., Takahara, Y., Theory of Hierarchical, Multilevel Systems, Academic Press, 1970.

/20/ Drew, Donald, Application of Multilevel Systems Theory to the Design of a Freeway Control System, Proc. Fifth Systems Symposium "Systems Approach and The City", Case Western Reserve University, 1970.

/21/ Cheneveaux, B., Synthesis of a Multivariable Control System Via Multilevel Techniques, 2nd IFAC Symposium on Multivariable Technical Control Systems, 1971.

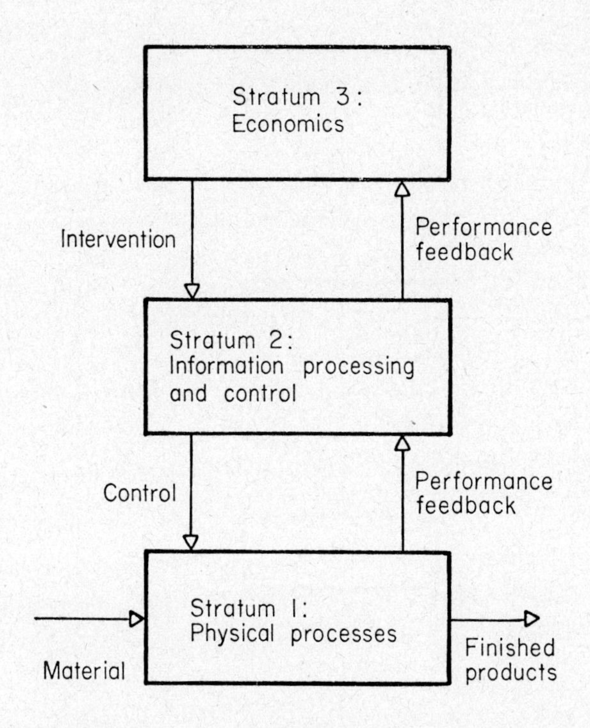
Stratum 3:
Economics
Intervention
Performance
feedback
Stratum 2:
Information processing
and control
Control
Performance
feedback
Stratum 1:
Physical processes
Material
Finished
products

Figure 1

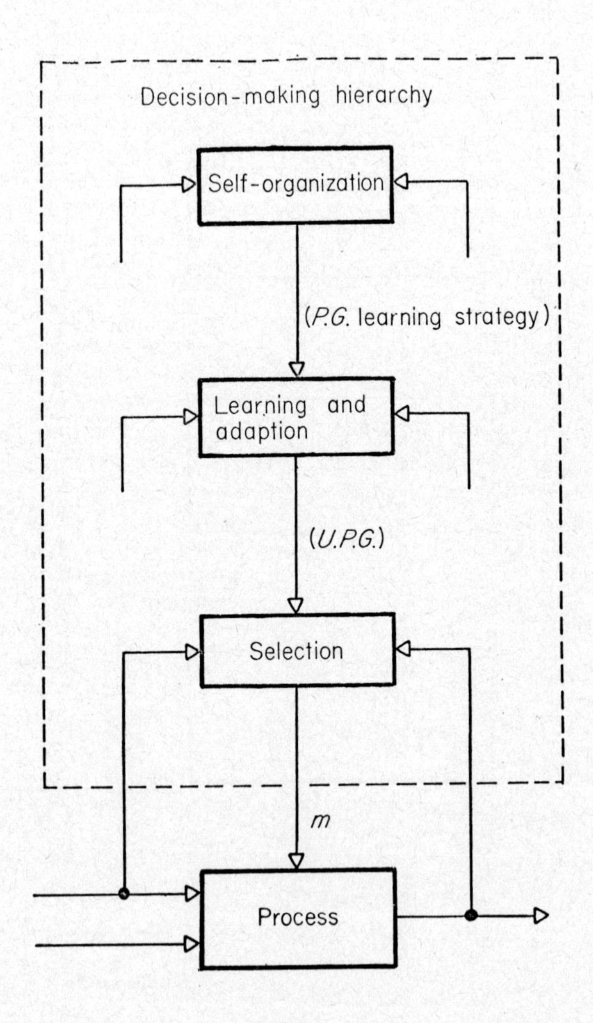
Decision-making hierarchy
Self-organization
(P.G. learning strategy)
Learning and adaption
(U.P.G.)
Selection
m
Process

Figure 2

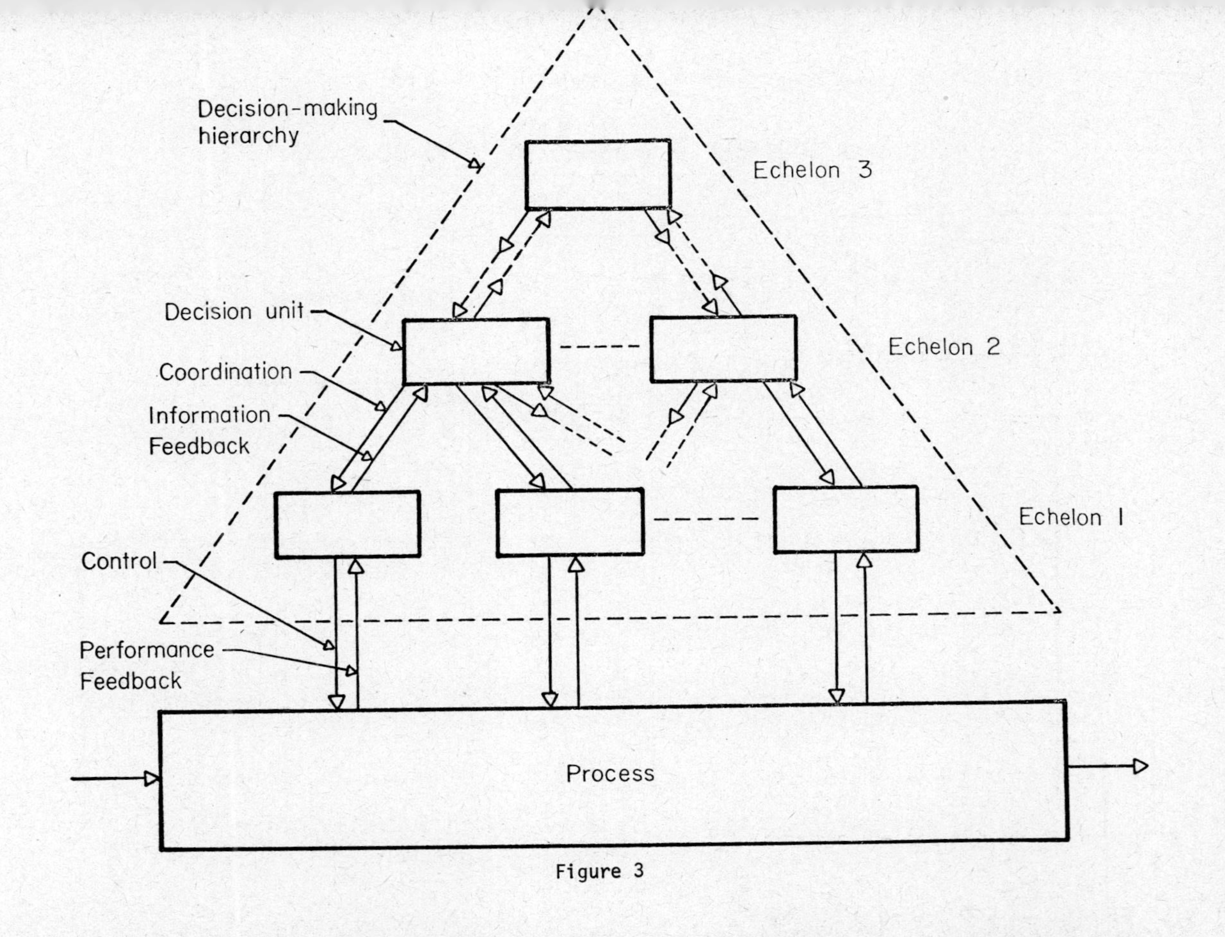
Decision-making
hierarchy
Echelon 3
Decision unit
Echelon 2
Coordination
Information
Feedback
Echelon I
Control
Performance
Feedback
Process

Figure 3

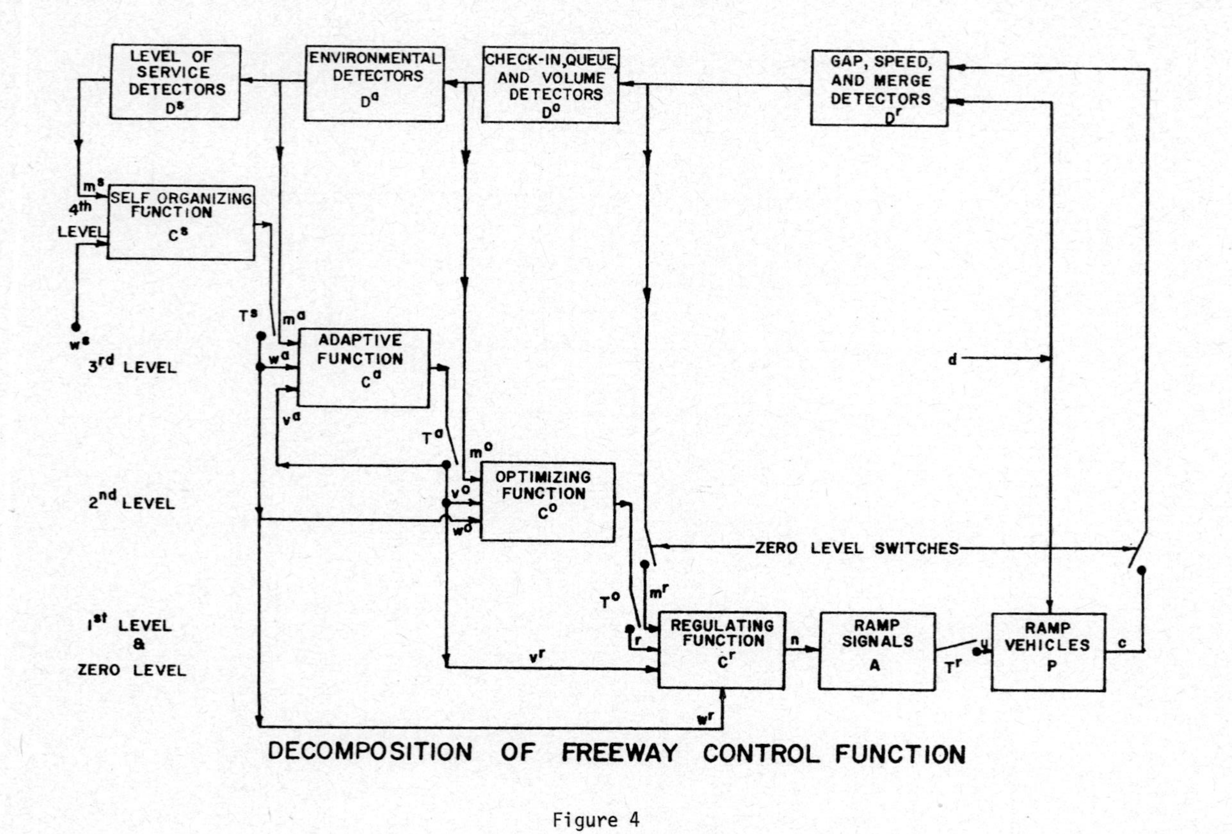

DECOMPOSITION OF FREEWAY CONTROL FUNCTION

Figure 4

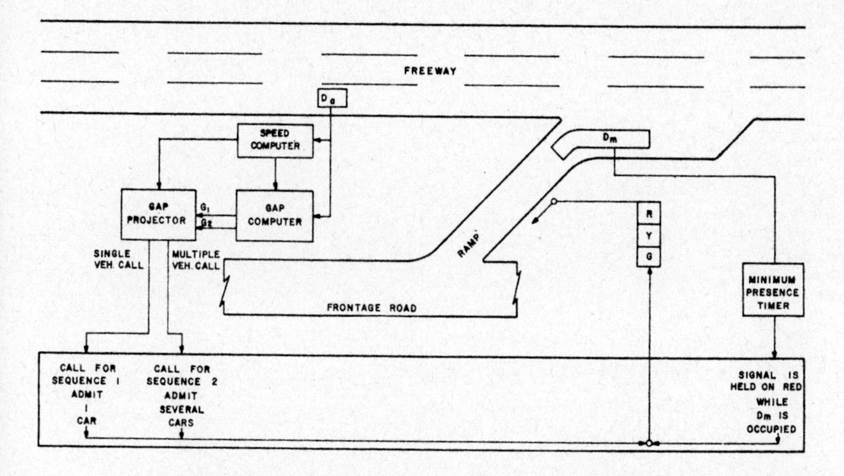

First level freeway control showing regulating function components.

Figure 5

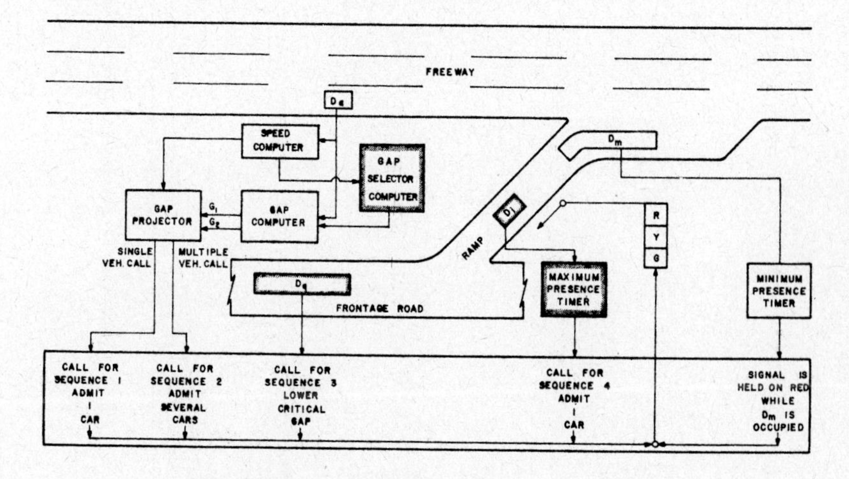

Second level freeway control with optimizing function components identified.

Figure 6

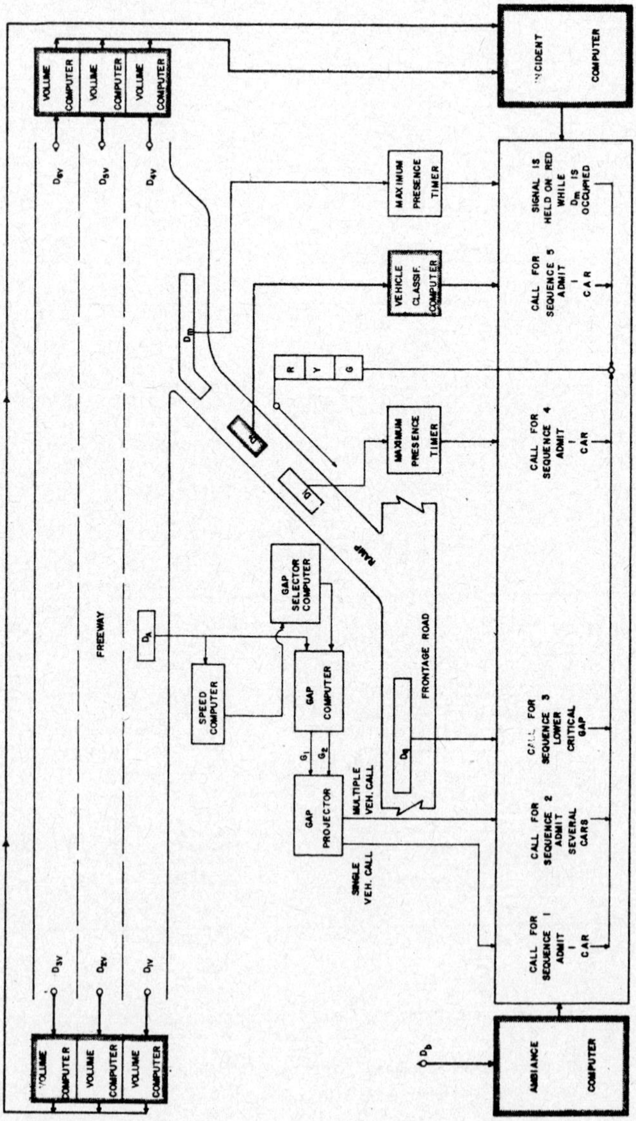

Third level freeway control with adaptive function components identified.

Figure 7

J.E. Rijnsdorp

MULTIVARIABLE SYSTEMS IN PROCESS CONTROL

MULTIVARIABLE SYSTEMS IN PROCESS CONTROL

J.E. Rijnsdorp
Prof.ir.
Twente University of Technology
Enschede / Netherlands

SUMMARY

The methods available for designing multivariable control systems are evaluated with respect to the requirements for process control. These methods have been combined in the following groups:
Analysis of interaction (The purpose is to obtain insight, usually for designing conventional control systems);
Invariance techniques (feedforward control, decoupling, non-interaction);
Modal and Pole-shifting techniques;
Optimum control techniques.
Their applicability depends strongly on the availability of good process models. Hence identification and adaptation techniques are also briefly discussed. Finally, something will be said about multi-level techniques in process control.

1. INTRODUCTION

Chemical processes have always been of a multivariable nature, whether this was taken into account or not. In the days when control theory only dealt with single controlled variables, the process control engineers were already faced with the problem of choosing the connections between controllers and correcting variables. As multivariable control theory was developed, this qualitative design problem obtained more and more quantitative overtones. Modern control concepts were being tried out by digital simulation, in laboratory or pilot plants, and, in some isolated cases, in commercial processes.

However, general application on an industrial scale is still around the corner, although the modern process control computer is fully capable of executing the required complex computations.

What is the stumbling block here? Is it the legendary mistrust of the process technologist and the process operator of anything theoretical? Is it a lack of salesmanship? Or is it perhaps a divergence between practical requirements and theoretical possibilities?

It is easier to formulate these questions than to answer them. In this paper, viewpoints from the literature will be given, supplemented with some personal ideas of the author.

1.1 Scope of the paper

The main subject of the paper will be simultaneous automatic control of two or more controlled variables by means of two or more correcting variables. Manual control will not be discussed, not because it is unimportant, but because little is known about manual control of slow processes. Something will be said about modelling and identification of multivariable processes, without going into details, however, since these are main subjects of other IFAC Symposia.

The term "process" will be taken in a wide sence; it encompasses not only the chemical but also the petroleum and the foodstuff industries, and it is firmly linked to steam generation for the production of electricity. In Section 2, more will be said about types of process models.

It is impossible to give a complete account of the literature. This survey will mainly be restricted to application papers, and to theoretical papers in chemical engineering journals. Before turning to the literature, a list of requirements for process control will be given.

1.2 Requirements for process control

Processes can be operated in three modes:
Steady operation. - , where the controlled variables should be kept near to given constant desired values.
Transient operation. - , where the controlled variables should be moved from an initial to a final state, preferably in an optimal way. This is typical for batch processes and for changing continuous processes over from one steady operation to another.
Cyclic operation. - Some processes run more efficiently if they are subjected to continuous variations, usually of a periodic nature (See / 1 - 11 /). As this subject is still mainly in the laboratory phase, and would require a lengthy discussion of process behaviour, it will not be discussed further.

The following requirements apply to steady operation:

1. The control system should keep the controlled variables near their desired values in spite of process disturbances and measurement noise. In many cases there is a constraint either above or below the desired value, which should only rarely be violated (See / 12 /). The presence of a constraint on one side makes the control problem asymmetric, hence it can be expected that symmetric control algorithms are nonoptimal.

2. The control system should satisfy requirement 1 for a variety of desired values corresponding to different throughputs, feed materials, product specifications, etc., and for different states of the process (degree of fouling, catalyst activity, process equipment out of operation, etc.).

This requirement asks for "homeostasis" (insensitiveness) to changes in process parameters, or for adaptive control.

3. The control system should also satisfy requirement 1 when one or more correcting variables have reached their constraints (control valves wide open or completely closed) or have been put on manual control. Only when the number of active correcting variables becomes too small, the least important desired values may be ignored.

4. The process operators should have the feeling that they know what the control system is doing, and be able to take over part or all of its functions in a smooth way.

5. The control system should not introduce such large and rapid variations into the correcting variables that other processes are unfavourably affected.

6. The control system should be linked to higher levels in the control hierarchy (See / 12-14 /), and Section 8).

7. The control system should minimise deviations and violations of constraints when large transients occur, such as caused by breakdown of process equipment. This points to the general principle of designing not only for stochastic avarage conditions but also for predefined severe conditions: An automobile designed only for the average speed (say 60 km/h) on a road with average slope (zero degrees) will fail completely on the "Autobahn" and in the mountains.

Finally, the following additional requirement applies to transient operation:

8. The process should be moved from one steady operation to another without violation of constraints and in an optimal way. Note that this requirement is typical for servomechanisms.

Obviously it is difficult to satisfy all these requirements with one control algorithm. Probably a combination of several techniques will be required, integrated in an adaptive structure.

2. PROCESS MODELS

As a good process model is a conditio sine qua non for application of control techniques, attention will be paid to 3 types of process models: qualitative, static, and dynamic ones.

2.1 Types of variables

Fig. 1 shows the input and output variables of a process as distinguished by Greenfield and Ward /15/ based on Bollinger and Lamb's /16 / analysis.

The following symbols are being used:
u: The vector of correcting variables;
v: The vector of measured independent variables (known disturbances;
w: The vector of non-measured independent variables (unknown disturbances);
c: The vector of controlled-and-measured variables;
m: The vector of measured but non-controlled output variables;
y: The vector of measured output variables (combination of c and m);
n: The vector of non-measured and non-controlled output variables;
x: The vector of state variables (usually the combination of y and n);
q: The vector of parameters in the process model.

The number of elements in the vectors will be denoted by dim(u), dim(v), etc.
Feedforward control can be realised by coupling u to v. However, feedback control, which is desirable in view of variations in w and q, can be realised by coupling u to c. The state x can be estimated from y, u and v.

2.2 Qualitative models

The great majority of control systems are still being designed on the basis of qualitative process models. Physical insight and experience indicate how to connect the controllers to the control valves. Also the choice of controller behaviour, which usually boils down to the adjustment of P-, I-, and D-action, is realised by trial and error. More sophisticated design techniques are only being applied in difficult cases, and whenever many similar systems are being designed (e.g. steam boilers).

2.3 Static models

Static models have been developed for a variety of processes, such as distillation columns, naphta crackers, thermal reformers, catalytic crackers, ammonia plants, etc./17-21/, (and T.J.Williams surveys in I & EC.) They consist of algebraic equations (usually nonlinear) and inequalities (corresponding to constraints). They are generally used for static optimisation of process operation, resulting in optimum desired values for the controlled variables. The objective function consists of product values and feed and utility costs. The optimum is determined by nonlinear programming. When the optimum is at one or more constraints pertaining to output variables, the method indicates that these variables should be controlled ("constraint control", see /22, 23/). In this way, the control scheme is partly determined by the optimisation. Of course, when the process conditions are changed, the optimum shifts to another location yielding other desired values and possibly other variables to be controlled.

An obvious drawback of static models is that dynamic effects are not to taken into account, but should be dealt with separately. However, dynamic models suitable for overall process optimisation are very complicated.

2.4 Dynamic models

It is not always appreciated that there are various types of dynamic process models, which differ widely from each other (see also Himmelblau and Bischoff /24 /). They are: Lumped models of low order, lumped models of high order, models with time-delays, and distributed-parameter models.

Low-order lumped models. - These models consist of a small number of differential equations. A common example is the continuously stirred tank reactor with ideal mixing and simple (often too simple) reaction kinetics, which, in view of its simplicity has been a real "guinea pig" for the application of modern control techniques.

It is generally easy to write the differential equations in state vector form. For chemical processes, where control valves influence flow rates, the result is often linear in u. Complete linearisation is permissible for steady operation (unless requirement 7 in Section 1.2 should be taken seriously into account).

High-order lumped models. - These models pertain to so-called multistage processes (e.g. distillation columns) or to multicomponent processes (e.g. polymerisation reactors). They consist of large sets of similar differential equations (often in the form of difference-differential equations), while the number of correcting and controlled variables still is relatively small.

Matrix A in the linearised state vector representation

$$x = A\,x + B\,u + C\,v + D\,w \qquad (1)$$

usually is rather empty; often it is of tri-diagonal form.

Of course, good order reduction methods are very valuable here. In some cases a careful study of the process responses can indicate a simple representation. For instance, the response of the tray concentrations in a tall distillation column to variations of the vapour flow can be approximated by an exponential lag with a very long time constant (See / 25 /). More systematic, general-purpose techniques can also be applied. For instance, Davison / 26 / and Aoki / 27 / have proposed methods whereby the most important eigenvalues of A are retained.

Models with Time Delays. - Models with time-delays are being used for processes with so-called plug flow (e.g. the idealised tubular chemical reactor; See /24,28 /), or as an approximation for multistage processes. The control theory of systems with time delays is presently in intensive development (See, e.g. the paper by Soliman and Ray in this Symposium).

Distributed-Parameter Models. - When the process variables are functions of time and of geometric coordinates, the dynamic behaviour must be described by partial differential equations. The control of distributed systems theory has already been developed to some extent. In a recent survey paper by Robinson /29/, however, only applications to aerospace and hydrodynamic problems have been mentioned. Robinson also indicates that the most popular single approach to distributed optimal control problems is to approximate these problems by some other problems which can be solved by more conventional techniques. In particular the method of transfer function approximation seems suitable for process control, because the number of input and output variables usually is relatively small. Approximation of calculated transfer functions by a relatively simple state vector model is therefore a good possibility of model building.

2.5 Canonical Structures

When the process model is formulated in terms of transfer functions, then different structures can exist. Mesarovic /30 / has defined two canonical structures: P- and V- type structures.

In the P-type every output variable is influenced by every input variable. In matrix notation (for c and u; see figure 1):

$$c(s) = P(s)u(s) \tag{2}$$

where P(s) is a dim(c) x dim(u) matrix of transfer functions.

In the V-type canonical structure every input variable influences only one output variable, but it is itself influenced by all other output variables. Here it is assumed that dim(c) = dim(u). In matrix notation (See also figure 2):

$$\begin{aligned} c(s) &= P^D_{cu}(s) \left(u(s) + P^N_{uc}(s)c(s) \right) \\ &= \left(I - P^D_{cu}(s)P^N_{uc}(s) \right)^{-1} P^D_{cu}(s)u(s) \end{aligned} \tag{3}$$

where the superscript D denotes a diagonal matrix, and the superscript N a non-diagonal matrix. The first subscript indicates the output, and the second the input.

Generally, differential equation and state vector models directly lead to a P-type structure. Subsequent transformation to Mesarovic's V-type structure can be realised by the following expressions (See also Schwarz / 31 /, p. 169-70)

$$p^{D}_{cu,ii}(s) = \det (P(s))/P_{ii}(s) ;$$

$$p^{N}_{uc,ij}(s) = -P_{ji}(s)/\det (P(s)) \qquad (4)$$

where lower-case letters represent matrix elements, and $P_{ji}(s)$ is the ji cofactor of P(s) (the transfer function matrix of the P-type structure). All transfer functions in P(s) are assumed to be stable.

Evidently, when det (P(s)) has non-minimum phase character (when it possesses one or more zeros with a positive real part), $P^{N}_{uc}(s)$ is unstable, and this V-type structure only has mathematical significance. This V-type and other related structures will be further discussed in Section 4.

3. ANALYSIS OF INTERACTION

The conventional approach to multivariable process control is to connect standard automatic controllers in such a way that the requirements are met as well as is possible. The controller transfer fuctions can be represented by a matrix in cascade with the transfer fuction matrix of the process. One obvious goal is to avoid severe interaction. Of course, the word "severe" deserves further qualification. It could mean that the controlled variables are mutually dependent, which impairs control at low frequencies. It could also mean that the system becomes unstable if one or more of the controllers are (taken) out of operation. A third interpretation of "severe" is the presence of non-minimum phase factors in the transfer functions as "seen" by the controllers.

In this Section various measures for these three types of severe interaction will be discussed. These measures can again be of two types: Those that only consist of process properties, and indicate the possibility of severe interaction if the process is controlled; and those that describe the complete system behaviour.

3.1 Weak Interaction

Before discussing severe interaction, we would like to know if the interaction is so weak that it may be neglected. This is particularly important for large processes, where decomposition into smaller process models is very desirable. An index based on the P-canonical structure is

$$J_1(\omega) = \frac{\Sigma \,|\, \text{PERMUTED TERMS OF } \det (P(j\omega)) \,|}{|\, \text{DIAGONAL TERM OF } \det (P(j\omega)) \,|} \qquad (5)$$

where the | | stand for modulus, and the determinant has been written as a sum of n! terms.

If $J_1(\omega) \ll 1$ for all frequencies, then the interaction is small, and the controlled system can be approximated by n single variable systems.

3.2 Dependency between Controlled Variables

If P is singular, the controlled variables form a dependent set. In practice one should avoid complete dependency by choosing correcting and controlled variables correctly. However, it can still occur that P is almost singular. A dimensionless measure for the ill-conditionedness is:

$$J_2(\omega) = \left| \frac{\det(P(j\omega))}{\prod_i p_{ii}(j\omega)} \right| \ll 1 \tag{6}$$

Formula 6 should only be evaluated for low frequencies, i.e. where control of the separate loops can be made efficient:

$$|p_{ii}(j\omega)c_{ii}(j\omega)| \gg 1 \tag{7}$$

For processes described in terms of state variables, Davison /99/ has used the degree of dependence of the columns of matrix B (See formula 1). He has applied this idea to control systems for distillation columns.

Mesarovic / 30 / has formulated a general type of interaction measure to indicate the relative sensitivities of output variables of the controlled system to a perturbation, e.g. a change in a parameter or transfer function (see also Schwarz / 31 /, p. 204-7).

However, these interaction measures are difficult to apply to the synthesis of control systems.

3.3 Immunity against Inoperative Loops

A necessary, but not sufficient condition for stability with one inoperative loop is (See Grabbe / 32 /).

$$J_3(0) = \frac{\det(P(0))}{\prod_i p_{ii}(0)} > 0 \tag{8}$$

The equivalent expressions, where any p_{ii} and the corresponding rows and columns of P have been omitted , should be used for investigating the stability with several inoperative loops. Rosenbrock / 33 / applies a theorem by Ostrowski to analyse the influence of variations of feedback controller gains between zero and a maximum value. The forward path of the multidimensional control loop is put into a matrix of transfer functions $Q(j\omega)$, the inverse of which is denoted as $\hat{Q}(j\omega)$.

A series of Nyquist diagrams is constructed of $\hat{q}_{ii}(j\omega)$, the the main diagonal elements of $\hat{Q}$ (See figure 3). Superimposed on the polar curve are interaction circles with radii equal to the sum of the moduli of the offdiagonal elements in row i of $\hat{Q}$. Rosenbrock (See / 34 /, p. 202) has shown that the closed loop system is stable if the inverse Nyquist criterion applies to the envelope of the superimposed circles. Apart from this result, the paper also proposes some interesting ideas for decoupling and for control system synthesis.

3.4 Non-minimum Phase Behaviour

The third interpretation of severe interaction is the generation of one or more non-minimum phase factors with long time constants (zeros with a small positive real parts) in the system response as "seen" by controller c_{ii}:

$$\frac{c_i(s)}{u_i(s)} = \frac{\det\ (\ P(s)\ )}{P_{ii}(s)} \tag{9}$$

Here it has been assumed that all controlled variables except c_i are ideally controlled, and a perturbation is introduced in u_i.

If det (P(s)) constains one or more non-minium phase factors with large time constants, then these will also appear in any control loop if the other loops are tightly controlled (See formula 7).

Davison / 35 / has developed a non-minimum phase index, which takes into account the presence of actual controllers (hence it is more precise than (9)). It is a function of the dominant pole of the system response, and of the zeros. Although the other poles do not appear, it should be checked if they have negative real parts, viz. if the controlled system is stable. This means that the amount of computation required to use the non-minimum phase index is hardly less than that required for calculating the exact system response.

Finally, Nisenfeld and Schultz / 36 / judge the stability of multivariable control systems by means of a matrix of ratios of controlled/uncontrolled responses. However, their analysis implicitely assumes pure dead-time dynamics, which is only realistic for a small number of processes (e.g. in-line blending).

The final conclusion about interactions measures is that those based on the process dynamics are relatively easy to compute but rather vague, while those based on controlled system dynamics are more precise but difficult to compute.

3.5 Two-variable Control Systems

Two-variable control systems can still be reasonably well analysed by classical single-variable methods. If they contain only two controllers (no "cross-controllers"), three loops can be distinguished: The two subloops with loop transfer functions $\Lambda_1 = P_{11}C_{11}$ and $\Lambda_2 = P_{22}C_{22}$, and the main (interaction) loop (See figure 4). The transfer function of the latter loop can be formulated in different ways depending on the way in which the loop is opened (See figure 4):

$$\kappa \frac{\Lambda_1}{1+\Lambda_1} \frac{\Lambda_2}{1+\Lambda_2} \tag{10}$$

$$\Lambda_i \left[1 - \kappa \frac{\Lambda_j}{1+\Lambda_j} \right] \tag{11}$$

$$\Lambda_i \frac{1+(1-\kappa)\Lambda_j}{1+\Lambda_j} \tag{12}$$

$$\Lambda_i \left[1+\Lambda_j/\Lambda_i \left[1+\Lambda_i(1-\kappa)\right] \right] \tag{13}$$

$$\text{where } \kappa = P_{12} P_{21} / P_{11} P_{22} \tag{14}$$

Note that κ is related to J_1 (See formula 5) and $(1-\kappa)$ to J_2 and J_3 (See formulae 6 and 8).

Various authors have used these formulae for analysing the interaction. Rijnsdorp / 37 / has used (12) in an inverse Nyquist diagram with an application to distillation columns. Sieler / 38/ has used (10) and (11) in Nichols diagrams with an application to pressure and flow control. Muckli has applied a modified form of (10). Schwarz / 39 / has used (13) in a sequence of root-locus diagrams.

Starkermann /40-42/ and Kraemer / 43 / have made a thorough analysis of two-variable control systems with process transfer functions of different order and different types of controller actions. These publications are useful for obtaining insight into the behaviour of two-variable control systems, and thereby for designing more or less conventional control systems.

3.6 Interaction between Many Similar Control Loops

In some processes there is interaction between many similar control loops, e.g. when many similar processes draw from a common supply or feed to a common take-off.

A system of the latter type has been investigated by Kruidering /44 /. He has found that the controller gains had to be reduced by a factor 2 due to the interaction.

In some cases the characteristic equation can be factorised, and each factor can be handled by single variable techniques (See /45 /).

4. INVARIANCE TECHNIQUES

The term "invariance" has been introduced by Soviet scientists (Kulebakin /46 /, Petrov /47 /, Meerov /48 /, and others) to denote control actions intended to give exact compensation. It encompasses feedforward control, non-interaction, and prescribed behaviour ("model following"). These techniques will now be discussed separately, and in combination.

4.1 Decoupling

Decoupling is a way to achieve non-interaction, i.e. to reduce the multivariable control system to a number of single-variable control systems. It has been introduced by Boksenbom and Hood / 49 / in possibly the first application **to a real system.** Non-interaction can be defined in different ways (See Schwarz / 31 /, p. 377). Here we shall restrict ourselves to diagonalising the transfer function matrix of the multidimensional feedback loop.

Engel / 50 / has shown that there are many possible locations for the decoupling elements (See figure 5). Moreover, the decoupling elements can have P- or V-type structure, consequentely the total number of possibilities is very large. His analysis leads to the conclusion to locate the decoupling transfer functions between controllers and process (3 in figure 5) because this makes them independent of the controller transfer fuctions. Moreover, as has already been found by Mesarovic / 30 /, choosing the decoupling structure different from that of the process yields simple expressions. For instance, for a P-type process the following simple V-type structure is found (See also figure 6), and Schwarz / 31 / p. 428 for the two-variable case):

$$n_{ii}^{N}(s) = 0 \; ; \; n_{ij}^{N}(s) = p_{ij}(s)/p_{ii}(s) \qquad (15)$$

The system is only physically realisable when the feedback loop in the decoupling structure is stable (See Plote /51/) This is determined by the zeros of the characteristic equation

$$\det\,(\,I+N^{N}(s)\,) = 0\;,\;\text{hence}\;\det\,(\,P(s)\,)/\Pi p_{ii}(s) = 0 \qquad (16)$$

For the usual case that the process transfer functions are stable,(16) indicates that det (P(s)) should be minimum phase.

If det (P(s)) is non-minimum phase, there are evidently many kinds of difficulties (See also Section 3). As pointed out by Rosenbrock /52 /, for a two-variable example, decoupling a process where det (P(s)) is non-minimum phase by means of a P-type structure yields two control loops with non minimum phase character. This is worse than control without decoupling, whereby one of the control loops can always be made "tight", and only the other loop is stuck with non-minimum phase character.

Fitzpatrick and Law /53 /, have formulated P-canonical decoupling for sampled-data systems, whereby they use the z-transform for decoupling only at the sampling instants, and the modified z-transform for decoupling for all moments of time. Many other possibilities of decoupling can be found in Schwarz's book /31 /.

4.2 Feedforward Control

Feedforward techniques have received much attention in the process control field. There are many papers about applications to distillation columns / 54 /, also in combination with optimisation /55,56/, to chemical reactors / 57,58 /, evaporators / 59 /, etc.

For a linear model, feedforward control is straightforward: it amounts to coupling the measured independent variables v (See figure 1) to the correcting variables u via a matrix of transfer functions $C_{uv}(s)$ such that (See figure 7):

$$P_{cv}(s) + P_{cu}(s)C_{uv}(s) = 0 \quad ,\text{or } C_{uv}(s) = -P_{cu}^{-1}(s)P_{cv}(s) \qquad (17)$$

where dim (u) = dim (c), and $P_{cu}(s)$ is non-singular.

$P_{cu}(s)$ should not be non-minimum phase (See Luyben /60 / for an example of the consequences). If dim(u)>dim(c) some of the elements of $C_{uv}(s)$ can be chosen freely (Singer / 61 /).

A more general type of feedforward is known as the principle of invariance (See beginning of Section 4). Here not only independent variables, but also state variables are used as inputs to the control matrix. For the general case of a nonlinear process model the derivatives of the controlled variables ($\dot{c}$) are set equal to zero, and their actual values (c) equal to the desired values. The resulting set of dim(c) algebraic equations and dim(x)-dim(c) differential equations can be solved for the dim(u) correcting variables, whereby a number of independent variables (measured of unmeasured) can also be eliminated.

The final result expresses the correcting variables as nonlinear (or linearised) functions of selected state and independent variables.

Luyben /62 / discusses an application of this idea to a model of an ideally mixed continuous tank reactor with dim(x)=dim(y)=dim(u)=dim(v) = 2. The final equations express u in terms of the desired values of y, and v. For an application to a real process, however the process model should be supplemented by instrumentation dynamics, which makes the system more complicated.

4.3 Decoupling and Feedforward control

The combination of decoupling feedforward control is analysed by Foster and Stevens /63 /. They use a linear modified V-type structure for the process, which contains dummy state variables in order to make the matrix square (See also Horowitz /64 /). This method is limited to cases where all state variables are being controlled (very rare in process control).

Greenfield and Ward / 65 / have extended this idea by further modifying the V-type structure, whereby the nonmeasured state variables are eliminated from the Laplace-transformed process model. In another article / 66 /, they had already worked out a more subtle form of this structure, which they call structural analysis. After elimination of n from the original state equations, they arrive at the following expression (See also figure 8).

$$(sI-Q^{D}_{cc}(s)) c(s) = Q^{D}_{cu}(s) \quad Q^{N}_{cu}(s) u(s)+Iu(s)+Q_{cv}(s) v(s)$$

$$+Q_{cm}(s) m(s)+Q^{N}_{cc}(s) c(s)+Q_{cw}(s) w(s) \qquad (18)$$

Application of the invariance principle leads to the following control equation:

$$u(s) =-Q^{N}_{cu}(s) u(s)-Q_{cv}(s) v(s)-Q_{cm}(s) m(s)-Q^{N}_{cc}(s) c(s) \qquad (19)$$

In the terminology of Greenfield and Ward, the first term represents feedforward decoupling, the second feedforward, the third and the fourth feedback decoupling. Substitution of (19) into (18) yields

$$(sI-Q^{D}_{cc}(s)) c(s) = Q^{D}_{cu}(s)Q_{cw}(s) w(s) \qquad (20)$$

which only contains the unmeasured input variables w.

The advantage of this method over previous ones is that the order of the polynomial matrices to be inverted is minimised, hence the control transfer functions are less complicated.

4.4 Decoupling and Prescribed Behaviour

A well known synthesis method (sometimes denoted by model following) is to prescribe the overall system behaviour, and subsequently find a control system which realises this behaviour. For instance, in decoupling one cannot only require that the open-loop matrix is diagonalised, but that the diagonal elements of the process are left unchanged:

$$P(s)K(s) = P^{D}(s) \quad , \text{ hence } K(s) = P^{-1}(s)P^{D}(s) \tag{21}$$

where superscript D indicates that all nondiagonal are made equal to zero. Luyben /67 / has applied this idea to top-and bottom-product composition control of a distillation column. The results are rather poor, probably because the interaction is very severe (P(s) is almost singular) in this case. Schwarz / 31 / has analysed many structural variants of this idea, and has derived conditions for realisability.

Föllinger / 68 / obtains design conditions by prescribing that the controlled variables return exactly to the desired values after a predetermined time for a step disturbance. Shean-Lin Liu / 69 / considers the nonlinear process model

$$\dot{x} = F(x, u, t) \tag{22}$$

and prescribes

$$\dot{x} = G(x - x_d, t) \tag{23}$$

where x_d is the desired value of x. Combining (22) and (23) yields

$$F(x, u, t) = G(x - x_d, t) \tag{24}$$

which can be solved for u. Evidently, this method only works if the number of state variables is not larger than the number of correcting variables. Still Shean-Lin Liu gives a calculated example for an absorber, a typical multistage process with dim(x)>>dim(u). He uses a truncated process model, which only consists of one end of the absorber. In the practical application, however, measurement dynamics and errors would cause great difficulties, particularly for compositions where the measuring time lags are much longer than the process lags of one stage.

4.5 General conclusions about Invariance Techniques

Feedforward control is only effective if an accurate process model is used. This is not required for decoupling since only severe interaction should be avoided. Both control techniques run into difficulties when the transfer function matrix of the process is non-minimum phase.

More elaborate invariance methods, whereby not only input but also output variables of the process are used, look promising for low-order lumped models.

5. MODAL ANALYSIS AND POLE-SHIFTING METHODS

5.1 Modal Analysis

In 1962, Rosenbrock / 70/ proposed a novel method for tackling multivariable control problems. The principle is to make the dominant transient modes (corresponding to the dominant eigenvalues of the system matrix A:

$$\dot{x} = A\,x + u \qquad (25)$$

less dominant, and simultaneously avoid interaction as much as possible.

A is for this purpose transformed into:

$$A = T\,J\,T^{-1} \qquad (26)$$

where J is the Jordan canonical form of A (if all eigenvalues of A are real and distinct, it is a diagonal matrix in these eigenvalues). T is the column-matrix of right-hand eigenvectors of A, and T^{-1} the row-matrix of left-hand eigenvectors. Now the following control law is applied:

$$u = -\,T\,K^{D}\,T^{-1}\,x \qquad (27)$$

Here T^{-1} selectively measures the various modes; the diagonal matrix K^D shifts the corresponding eigenvalues to more negative values, and T ensures that every output of K^D only affects the corresponding mode. This can be seen from the combination of (27) and (28):

$$\dot{x} = T\,(\,J - K^{D}\,)\,T^{-1}\,x \qquad (28)$$

However, this principle cannot be applied as such in practice because usually dim(u)<dim(x) (Compare formula 25 to formula 1) and dim(y)<dim(x). These restrictions have been partly accounted for by Rosenbrock / 70 /.

More recently, Wonham / 71 / has developed a more general form of model analysis, the so-called pole assignment method, whereby the number of modes that can be influenced is equal to max(dim(u), dim(y)) (See Davison and Chatterjee / 72 /).

Modal analysis has been applied to dynamic models of boilers (See Ellis and White /73 /, and Davison and Goldberg / 74 /, and of a distillation column (Davison / 75 /). Compared to conventional control, however, the improvement has not been very impressive thus far. This could be related to the fact that process transfer functions often have only one dominant eigenvalue, which can satisfactorily be reduced by one controller. Moreover, process transfer functions with sereral dominant eigenvalues can often be handled by multiloop (cascade) control schemes.

5.2 Transformation of the Matrix of Process Transfer Functions

The matrix of process transfer functions belongs to the class of rational matrices. In his recent book / 34 /, Rosenbrock has extended the theory of these matrices, and indicated applications to control system design (See also /52,76/).

A discussion of this approach is outside the scope of this survey paper. It goes without saying that an investigation of the applicability to process control is very worthwhile.

6. OPTIMAL CONTROL

In Section 1.2 a distinction has been made between steady operation and transient operation. For optimising steady operation a linearised process model seems satisfactory during most of the time (although it should be updated when conditions change; see requirements 2 and 3 in Section 1.2). However, large transients occur occasionally, which might require a nonlinear process model (requirement 7). The latter is unavoidable for transient operation requirement 8).

These various cases will be discussed in the following sections.

6.1 Steady Operation; Linear Process Model

When the objective function is a quadratic functional of the correcting (requirement 5, Section 1.2) and the controlled variables, it is well known that the optimal control can be formulated in the form of a linear state feedback law (See Anderson and Moore / 77 /, and Kwakernaak and Sivan / 78 /). As on-line application almost certainly requires a digital computer, the discrete-time version is most appropriate:

process model: $x\,(k+1) = A^{*}\,x(k) + B^{*}\,u(k)$ (29)

objective function: $J = \overline{x^T Q x} + \overline{u^T R u}$ (30)

where ——— indicates averaging with respect to k.

control law: $u(k) = - K x(k)$ (31)

where K is the limit of the solution of a set of recurrence relationships (to be recomputed whenever a large change occurs in the process).

Apparently, the control law only has proportional action. This is insufficient for process control since sustained deviations should be eliminated. As indicated by Yen-Ping Shih / 79 / for the single variable case, and extended by Newell and Fisher / 80 / to the multi-variable case, this problem can be solved by simply augmenting the state vector by the integrals of the controlled variables and by introducing the appropriate terms into the objective function. A more complicated solution has been proposed by Anderson and Moore / 77 /).

Newell and Fisher have applied this type of control to a pilot plant evaporator with dim(u)=3, dim(y)=3, and dim(x)=5+3. The results were very favourable compared to conventional control. Although the process is markedly nonlinear, control was effective for large disturbances.

A problem in the implementation is the estimation of the state x(k). Newell and Fisher were able to do this in a rather simple way, but in other cases (particularly for multistage processes; see Section 2.4), state estimation becomes rather cumbersome and the controller is of high order. Amongst others, Kwakernaak and Sivan / 81 / have indicated a suboptimal solution to this problem.

A worthwhile generalisation of the approach would be to combine feedback with feedforward control. The process model then corresponds to (compare to <u>formula 29</u>):

$$x(k+1) = A^* x(k) + B^* u(k) + C^* v(k) \quad (32)$$

and the control law:

$$u(k) = - K_1 x(k) - K_2 v(k) \quad (33)$$

Solheim and Saethre / 82 / have solved this problem by a Lagrangian multiplier technique, and applied it to a discretised model of a heat-exchanger. West and McGuire / 83 / have applied Merriam's parametric expansion technique. Koppel / 84 / has incorporated the disturbance vector into the state vector.

Processes with pure time delays can be reduced to the abovementioned formulation when the controller is preceded by a predictor (See Koppel / 84 /, and West and McGuire / 83 /). Another publication about optimal control of dead-time processes is by Ray / 85 /.

6.2 Steady Operation; Non-Linear Process Model

The linear state feedback approach can be extended to non-linear process models. This leads to a nonlinear Kalman filter for state estimation followed by a controller. Two recent papers by Weber and Lapidus / 86 / compare different methods for generating a suboptimal controller of the form:

$$u(k) = - K(k)\ x(k) \tag{34}$$

They also show an application to an absorber (dim(x)=6 , dim(u)=2).

Seinfeld / 87 / (See also Seinfeld c.s./88,89/) takes an instantaneous objective function and gives an application to the model of a continuously stirred tank reactor. The amount of calculation is quite large, which evidently is the price to be paid for using a more accurate process model.

6.3 Return to Steady Operation; Non-Linear Process Model

Quite a different problem steady operation is the optimal return to the desired values after a large disturbance has occurred (requirement 7 in Section 1.2). This is closely related to transient operation: going from one steady-state to another (requirement 8).

Gould, Evans and Kurihara / 90 / have investigated optimal return to the steady state for a simplified model of a fluid catalytic cracker (dim(x)=2, dim(u) =2). The optimum paths from several different starting points towards steady-state have been plotted in the state plane. Along each path there is a series of values u_1 and u_2, which can therefore be read off as functions of the state:

$$u_1 = f_1 (x_1 , x_2) \ ; \ u_2 = f_2 (x_1 , x_2) \tag{35}$$

This is a nonlinear feedback control law.

Nonlinear process models and objective functions often still are linear in the correcting variables. This means that optimum control is of the bang-bang type (in practice "soft" bangs are advisable in order to reduce disturbances in other processes). Douglas / 91 / has worked this out for a continuously stirred tank reactor with minimum time for returning to the steady state. He also offers the interesting suggestion to combine the nonlinear control for large deviations with a linear control law for small deviations.

This could be a good solution for the modelling problem: A detailed linear model for small deviations and a crude nonlinear model for large deviations, instead of a detailed nonlinear model for all deviations.

Pollard and Sargent / 92 / have investigated the optimum switch-over of distillation column, comparing different numerical methods for finding the control path. They were somewhat dismayed by the amount of computation involved. Davison and Monro / 93 / have applied a hill-climbing technique to the determination of the switching instants for bang-bang control.

6.4 Conclusions about Optimal Control

Optimal control is often interpreted as striving for minimum cost or maximum profit. In real processes, however, this is an unrealistic goal; the aim should rather to be improve control. For the multivariable case one would like to design multidimensional feedback and feedforward laws, which are in line with the requirements listed in Section 1.2.

The contribution of optimal control methods here is reduction of the number of design parameters. In fact, choosing many controller gains is replaced by choosing a smaller number of weighing factors in the objective function. Moreover, the latter are also more directly related to the actual performance of control (See e.g. Tyler and Tuteur / 94 /).

As long as deviations are small, a linear control law seems appropriate. When deviations become large, control can be transferred to a suboptimal nonlinear control law (either continuous, see e.g. Gould c.s. / 90 /, or quasi bang-bang, see e.g. Douglas / 91 /).

7. STATE ESTIMATION AND MODEL IDENTIFICATION

It has already been said in Section 1.1 that model identification is essential for the realisation of improved control. In some cases a simple model of the correct order can be formulated. Then model identification reduces to parameter estimation. In other cases (e.g. for multistage or distributed processes), however, the order is not known a priori. This poses a difficult problem, particularly for on-line applications /95, 96/.

For optimum control, estimation of the state is usually required too. It can be expected that the difference between small and large deviations (See Section 6) is very important here. In the former case a linear model is sufficient, which can continuously be updated. In the latter case a nonlinear model should be used, which can only be updated when large distubances occur.

Recent information about system identification can be found in Åström and Eykhoff's survey paper /97/.

8. CONTROL HIERARCHY

In the chemical industry, process control represents the lowest "stratum" (in the terminology of Mesarovic c.s., see /98/) in a multi-level system (12-14). The next "stratum" determines the desired values for the controlled variables of the process (See figure 9), preferably based on optimisation of an economic criterion. The latter is determined by the "factory stratum", which has the task of coordinating the operation of the various processes in a given period of time (operations planning) in the best time sequence (operations scheduling).

The need for "layering" (See /98/) in the process control "stratum" has already become apparent in the discussion of the requirements for process control (See end of Section 1.2). In Figure 9 this is indicated by the block "control adaptation", which influences controller structure and parameters based on model identification (block "process models") and process optimisation. The integration of identification, optimisation, adaptation, planning and scheduling techniques poses a challenging problem for future work in the field of multivariable systems in process control.

Acknowledgment

The author wishes to thank prof.dr.ir. H. Kwakernaak for his valuable critical remarks and suggestions.

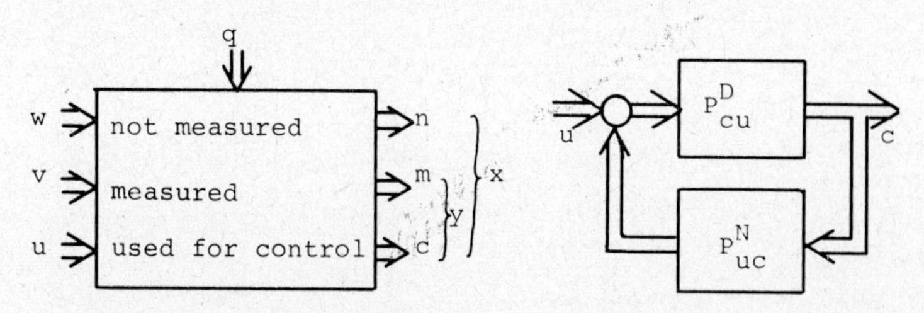

Figure 1. Process input and output variables

Figure 2. V-canonical structure

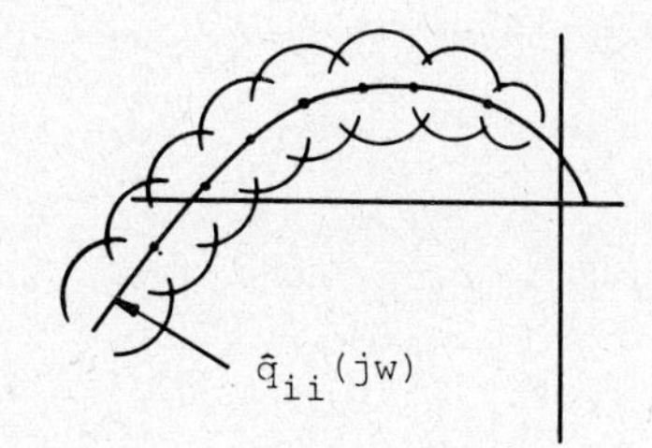

Figure 3. Inverse Nyquist diagram used by Rosenbrock

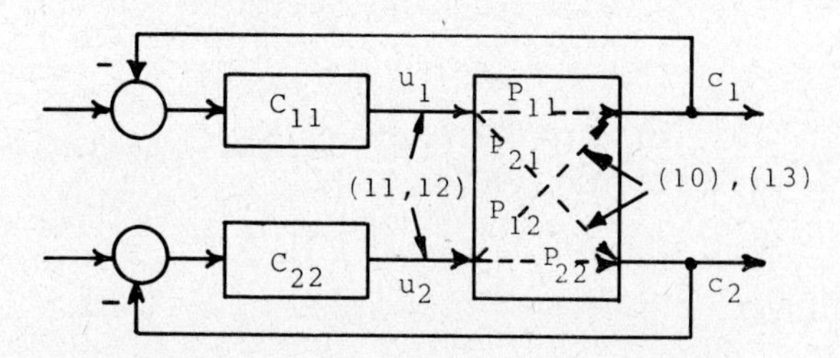

Figure 4. Two-variable control system.

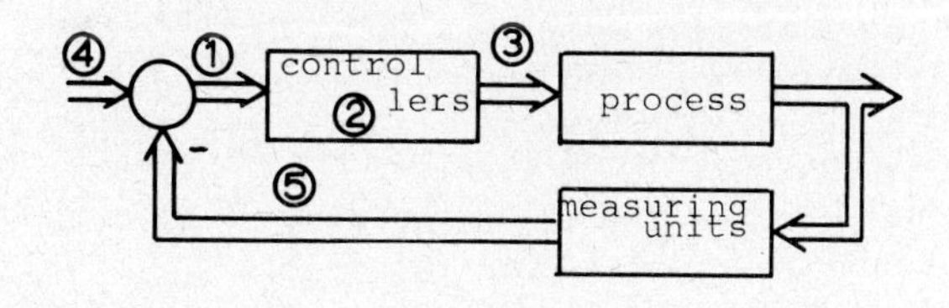

Figure 5. Possible locations for the decoupling elements.

REFERENCES

/1/ Cannon, M.R.: I&EC 53 (1961) No.8, 629-634

/2/ Robertson, D.C., Engel, A.J.: I&EC Proc. Des. & Dev. 6 (1967) No.1, 2-6

/3/ Gore, F.E.: I&EC Proc. Des. & Dev. 6 (1967) No.1, 10-16

/4/ Horn, F.J.M., Lin, R.C.: I&EC Proc. Des. & Dev. 6 (1967) No.1, 21-30

/5/ Horn, F.J.M.: I&EC Proc. Des. & Dev. 6 (1967) No.1, 30-35

/6/ May, R.A., Horn, F.J.M.: I&EC Chem. Proc. Des. & Dev. 7 (1968) No.1, 61-64

/7/ Belter, P.A., Speaker, S.M.: I&EC Proc. Des. & Dev. 6 (1967) No.1, 36-42

/8/ Douglas, J.M.: I&EC Proc. Des. & Dev. 6 (1967) No.1, 43-48

/9/ Schrodt, V.N., Sommerfeld, J.T., Martin, O.R., Parisot, P.E., Chien, H.H.: Chem. Eng. Sci. 22 (1967) 759-768

/10/ Chang, K.S., Bankoff, S.G,: I&EC Fund.7 (1968) No.4, 633-639

/11/ Fjeld, M.: Automatica 5 (1969) No.4, 497-506

/12/ Rijnsdorp, J.E.: Chem. Eng. Progr. 63 (1967) No.7, 97-116

/13/ Jong, J.J. de, Landstra, J.A., Rijnsdorp, J.E., Timmers, A.C.: Proc. 2nd IFAC Congress, Basle Butterworth / Oldenbourg (1964) 222-228, Butterworth London (1963)

/14/ Jong, J.J. de: Proc. 3rd IFAC Congress, London, 1966 paper 2A,Inst. Mech. Engrs.

/15/ Greenfield, G.G., Ward, T.J.: I&EC Fund. 6 (1967) No.4, 564-571

/16/ Bollinger, R.E., Lamb. D.E.: I&EC Fund 1 (1962) No.4, 245-252

/17/ Schöne, A.: Prozeszrechensystemen der Verfahrensindustrie, Hanser (München, 1969)

/18/ Proc. 1st IFAC/IFIP Symposium on Digital Computer Applications to Process Control, Stockholm (1964)

/19/ Proc. 2nd IFAC/IFIP Symposium on Digital Computer Applications to Process Control, Menton (1967), ISA

/20/ Preprints, 3rd IFAC/IFIP Symposium on Dig. Comp. Appl. to Proc. Contr., Helsinki (1971)

/21/ Grinten, P.M.E.M. van der, Buys, J.J.H.: Proc. Intl Symp. Distillation (1969) Brighton, Inst. Chem. Engrs., 6:13 - 6:21

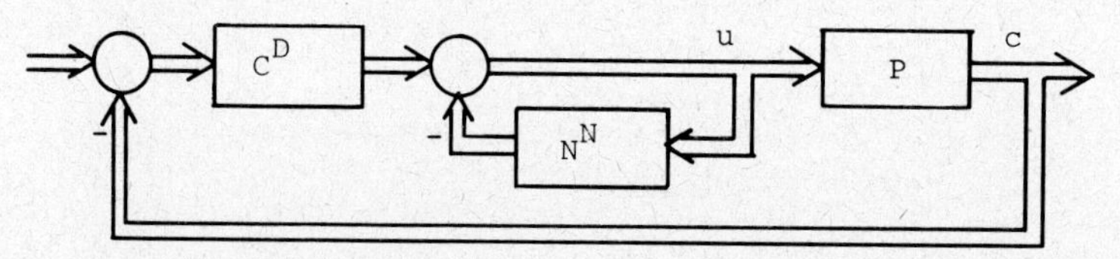

Figure 6. Simple decoupling of P-canonical process.

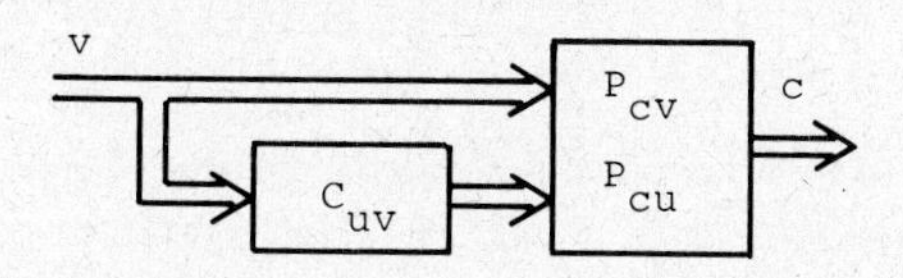

Figure 7. Feedforward control

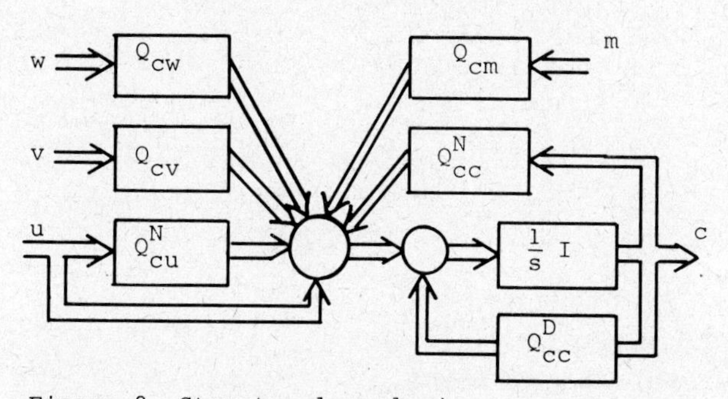

Figure 8. Structural analysis

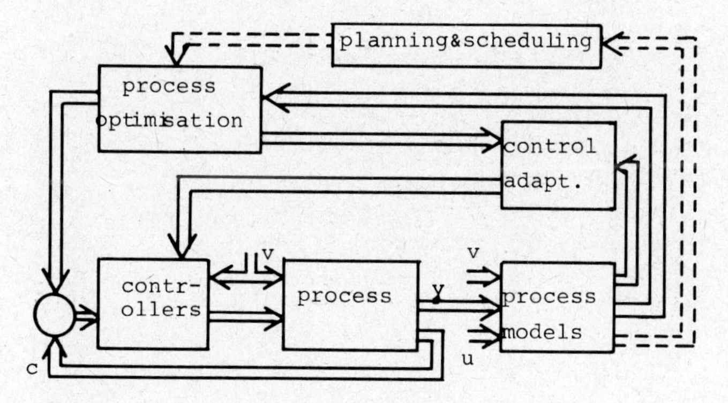

Figure 9. Control hierarchy

/22/ Maarleveld, A., Rijnsdorp, J.E.: Automatica 6 (1970) No.6, 51-58

/23/ Rijnsdorp, J.E., Maarleveld, A.: Proc. Intl Symp. Distillation (1969) Brighton,Inst. Chem. Engrs., 6:33 - 6:38

/24/ Himmelblau, D.M., Bischoff, K.B.: Process Analysis and Simulation, Deterministic Systems, Wiley (1968)

/25/ Wahl, E.F., Harriott, P.: I&EC Proc. Des. & Dev. 9 (1970) 396-407

/26/ Davison, E.J.: Tr. IEEE AC-11 (1966) No.1, 93-101

/27/ Aoki, M.: Tr. IEEE AC-13 (1968) No.3, 246-253

/28/ Ray, W.H., Soliman, M.A.: Chem. Eng. Sci. 25 (1970) 1911

/29/ Robinson, A.C.: Automatica 7 (1971) No.3, 371-388

/30/ Mesarovic, M.D.: The Control of Multivariable Systems, Wiley (1960)

/31/ Schwarz, H.: Mehrfachregelungen Band I, Springer (1967)

/32/ Grabbe, E.M., Ramo, S., Wooldordge, D.E.: Handbook of Automation Computation and Control, Vol 3 (1961), Chapter 10

/33/ Rosenbrock, H.H.: Meas. & Contr. 4 (1971) No.1, 9-11

/34/ Rosenbrock, H.H.: State-space and Multivariable Theory, Nelson (1970)

/35/ Davison, E.J.: Automatica 5 (1969), 791-9

/36/ Nisenfeld, A.E., Schultz, H.M.: ISA Natl. Conv. Philadelphia (Oct. 1970)

/37/ Rijnsdorp, J.E.: Automatica 1 (1965) No.1, 15-52

/38/ Sieler, W.: VDI-Lehrgang, Stuttgart (March 1966)

/39/ Schwarz, H.: Regelungstechnik 15 (1967) No.6, 257-262

/40/ Starkermann, R.: Regelungstechnik 16 (1968) No.7, 295-302

/41/ Starkermann, R.: Neue Technik 2 (1960) No. 4, 24-30

/42/ Starkermann, R.: Regelungstechnik 6 (1964), 242-249

/43/ Kraemer, W.: Fortschr. Ber. VDI-Z 8 (1968) No.10

/44/ Kruidering, B.C.: 1st IFAC Symp. Multivariable Control Systems, Düsseldorf (1967)

/45/ Bohn, E.V., Kasvand, Kasvand, T.: Proc. IEE 110 (1963) No.5, 989-997

/46/ Kulebakin, V.S.: Proc. 1st IFAC Congress (1960) Moscow, Butterworths (1961), 106-109

/47/ Petrov, B.N.: Proc. 1st IFAC Congress (1960), Moscow, Butterworths (1961), 117-122

/48/ Meerov, M.V.: Multivariable Control Systems, Moskow (1965)

/49/ Boksenborm, A.S., Hood, R.: N.A.C.A. Techn. Rep. 980 (1950)

/50/ Engel, W.: Regelungstechnik 14 (1966) No.12, 562-568

/51/ Plote, E.: Diplomarbeid, TH Hannover (1966)

/52/ Rosenbrock, H.H.: Proc. 3rd IFAC Congress (London) 1966, Paper 1A, Inst. Mech. Engrs.

/53/ Fitzpatrick, T.J., Law, V.J.: Chem. Eng. Sci. 25 (1970), 867-873

/54/ Luyben, W.L., Gerster, J.A.: I&EC Proc. Des. Dev. 3 (1964) No.4, 374-381

/55/ Lupfer, D.E., Parsons, J.R.: Chem. Eng. Progr. 58 (1962) No. 9, 37-42

/56/ Lupfer, D.E., Johnson, M.L.: I.S.A. Trans. 3 (1964) No.2 (Apr.), 165-174

/57/ Harris, J.T., Schechter, R.S.: I&EC Proc. Des. Dev. 2 (1963) No.3, 245-252

/58/ Tinkler, J.D., Lamb, D.E.: Chem. Eng. Progr. Symp. Ser. 61 (1965) No.55, 155-167

/59/ Nisenfeld, A.E., Hoyle, D.L.: Instr. Techn. 17 (1970) No.2, 49-54

/60/ Luyben, W.L.: Proc. Intl Symp. Distillation (1969) Brighton, Inst. Chem. Engrs., 6:39 - 6:48

/61/ Singer, D.: Regelungstechnik 15 (1967) No.9, 400-404

/62/ Luyben, W.L.: A.I.Ch.E.J. 14 (1968) No.1, 37-45

/63/ Foster, R.D., Stevens, W.F.: A.I.Ch.E.J. 13 (1967) No.2, 334-345

/64/ Horowitz, I.M.: Synthesis of Feedback Systems, Academic Press (1963)

/65/ Greenfield, G.G., Ward, T.J.: A.I.Ch.E.J. 14 (1968) No.5, 783-789

/66/ Greenfield, G.G., Ward, T.J.: I&EC Fund 6 (1967) No.4, 571-580

/67/ Luyben, W.L.: A.I.Ch.E.J. 16 (1970) No.2, 198-203

/68/ Föllinger, O.: Regelungstechnik 16 (1968) No.10, 449-454

/69/ Shean-Lin Liu: I&EC 6 (1967) No.4, 460-468

/70/ Rosenbrock, H.H.: Chem. Eng. Progr. 58 (1962) No.9 43-50

/71/ Morse, A.S., Wonham, W.M.: SIAM J. Control 8 (1970) No.3, 317-337

/72/ Davison, E.J., Chatterjee, R.: IEEE AC-16 (1971) No.1, 98-99

/73/ Ellis, J.K., White, G.W.T.: Control (1965) (April) 193-197, (May), 262-266

/74/ Davison, E.J., Goldberg, R.W.: Automatica 5 (1969) No.3, 335-346

/75/ Davison, E.J.: Trans. Inst. Chem. Engrs. 45 (1967) No.6. T229-T250

/76/ Daneels, A.: Revue A 9 (1967) No.2

/77/ Anderson, B.D.O., Moore, J.B.: Linear Optimal Control, Prentice-Hall (1970)

/78/ Kwakernaak, H.,Sivan,R:Linear Optimal Control Systems, Wiley-Interscience (to appear 1972)

/79/ Yen-Ping Shih: I&EC Fund. 9 (1970) No.1, 35

/80/ Newell, R.B., Fisher, D.G.: Preprint 3rd IFAC/IFIP Symp. on Dig. Comp. Appl. to Proc. Contr., Helsinki (1971)

/81/ Kwakernaak, H., Sivan, R.: Princeton Conf. on Int. Sci. and Syst. (March 1971)

/82/ Solheim, O., Saethre, A.: IFAC Symp. on Multivariable Control Systems, Dusseldorf (1968)

/83/ West, H.H., McGuire, M.L.: I&EC Fund. 8 (1969) No.2, 253-257

/84/ Koppel, L.: Introduction to Control Theory, Prentice-Hall (1968)

/85/ Ray, W.H.: Chem. Eng. Sci. 24 (1969), 209-216

/86/ Weber, A.P.J., Lapidus, L.: A.I.Ch.E.J. 17 (1971) No.3, 641-658

/87/ Seinfeld, J.H.: A.I.Ch.E.J. 16 (1970) No.11, 1016-1022

/88/ Seinfeld, J.H., Gavalas, G.R., Hwang, M.: Ind. Chem. Fund. 8 (1969) No.2, 257-262

/89/ Gavalas, G.R., Seinfeld, J.H.: Chem. Eng. Sci. 24 (1969) 625-636

/90/ Gould, L.A., Evans, L.B., Kurihara, H.: Automatica 6 (1970) No.5, 695-704

/91/ Douglas, J.M.: Chem. Eng. Sci. 21 (1966) 519-532

/92/ Pollard, G.P., Sargent, R.W.H.: Automatica 6 (1970) No.1, 59-76

/93/ Davison, E.J., Monro, D.M.: Automatica 7 (1971), No.2, 255-260

/94/ Tyler, J.S., Tuteur, F.B.: Tr.IEEE AC-11 (1966) No.1, 84-92

/95/ Isermann, R.: Automatica 7 (1971) No.2, 191-197

/96/ Gustavsson, I.: 2nd IFAC Symp. on Identification and Proc. Par. Est., Prague (1970)

/97/ Åström, K.J., Eykhoff, P.: Automatica 7 (1971) No.2, 123-162

/98/ Mesarovic, M.D., Macko, D., Takahara, Y.: Theory of Hierarchical, Multilevel Systems, Academic Press (1970)

/99/ Davison, E.J.: Automatica 6 (1970) 447-461

J.P. Waha

MULTIVARIABLE TECHNICAL CONTROL SYSTEMS. SURVEY OF APPLICATIONS IN POWER PLANTS AND POWER DISTRIBUTION SYSTEMS

MULTIVARIABLE TECHNICAL CONTROL SYSTEMS.
SURVEY OF APPLICATIONS IN POWER PLANTS AND POWER DISTRIBUTION SYSTEMS.

J.P. WAHA
S.A. Intercom
Brussels

SUMMARY.

Electric power generation, transport and distribution is a complex process, naturally interactive.

At the power-plant level where the aim is to deliver power to the network with given availability, quality and cost characteristics, problems of load control and start-up phenomenons are multivariable in numerous aspects.

Network operation based on decentralized logic (relay operation) and centralized supervision and control (power-frequency control, security assessment), is again multivariable.

Such a process is naturally organized in a multilevel way, direct and optimal control being shared amongst levels through coordination organization.

On-line control decentralized at plant and network level, and centralized at dispatch level using nowadays computer systems, is rended more intricate than before.

The techniques needed to build such an integrated control systems include modelling, multivariable control, optimization algorithms, -, and new technological development both in hardware and software of computers and control devices.

MULTIVARIABLE TECHNICAL CONTROL SYSTEMS.

SURVEY OF APPLICATIONS IN POWER PLANTS AND POWER DISTRIBUTION SYSTEMS.

J.P. WAHA
S.A. Intercom
Brussels

- - - - - - -

1. INTRODUCTION.

1.1. Some multivariable aspects of the operation of power plants and distribution networks are given in this survey.

This paper is organized in the following way :

- first a survey of identification and modelling techniques of multivariable processes is made,

- secondly some control problems are evoked, in a way where it is shown how high dimensional problems are solved in an industrial context

- next, a brief description of optimal control techniques as applied in power industry is done, the purpose being to analyse some mathematical programming techniques proposed at present times.

1.2. The return of the systematic use of control techniques in the power industry is nowadays well known; /1/ describes for instance the security, quality and economicity aspects of these techniques in the operation of plants and networks.

Nevertheless multivariable control techniques are not yet integrated as a whole in the resources every control engineer has to solve the problems which arise in this industry.

It is worthwhile to stress upon some aspects of operation of plants and networks with regards of the above criteria of security, quality and economicity. The operating people are faced with a high dimensional control problem to insure that the equilibrium between electricity production and consumption is made, while being sure that the system is stable, secure (no overloads of network elements), furnishes power of required quality (frequency, voltage), and operates at minimum cost.

Fig 1 sketches the overall process, production of power, transport and distribution network being shown. To operate such an integrated process consists for the control engineer to respect quality and security constraints regarding some state variables, or function of them /2/ of this process, while optimising a given economic criteria.

At the production level, fig 2 gives some of these variables for a classic thermal set, i.e. boiler outlet steam pressures, several

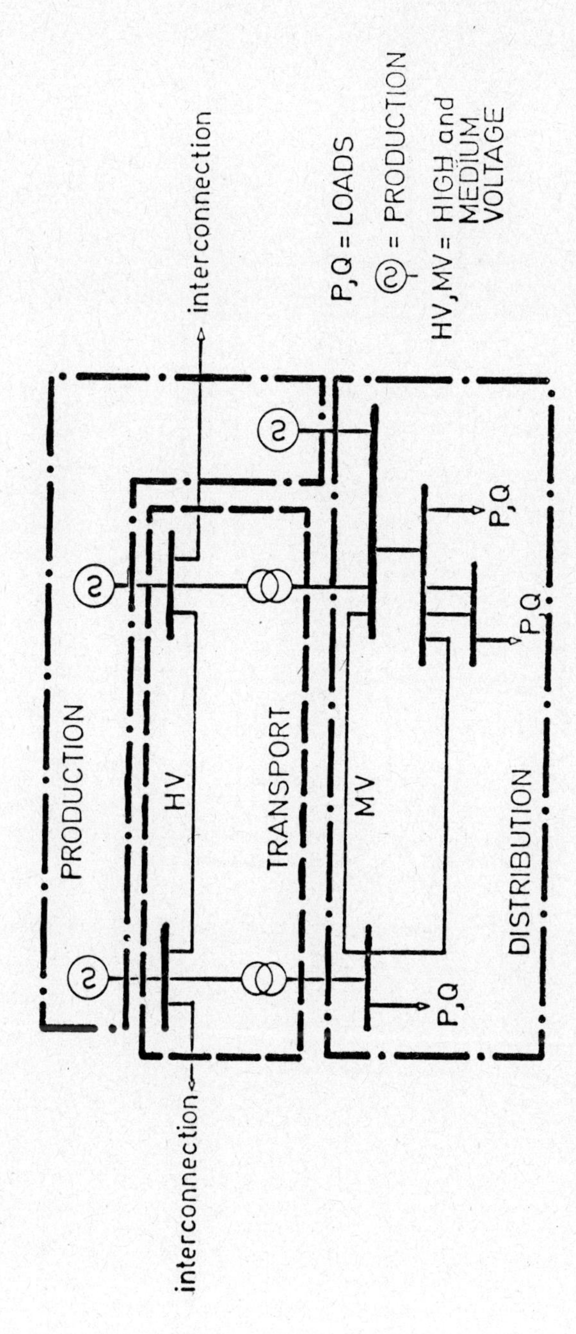

FIG.1 Hierarchical structure of the power industry

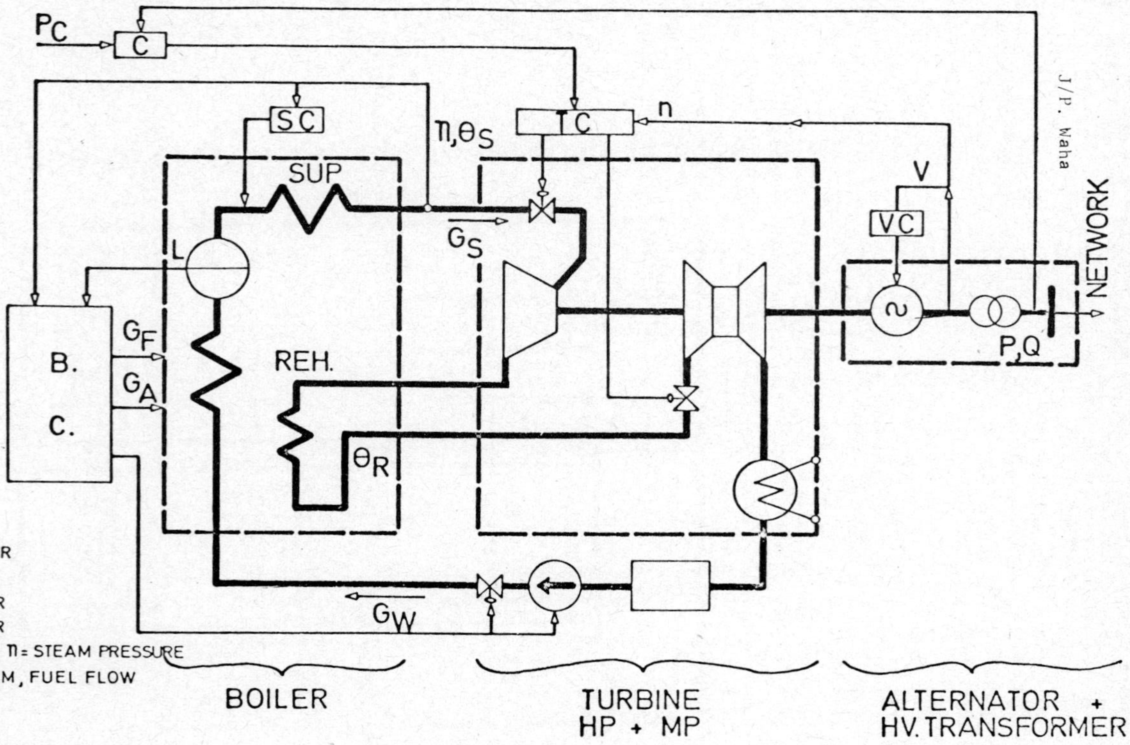

P_C C SC TC n Π, θ_S V VC SUP L G_S B. C. G_F G_A REH. θ_R G_W P,Q NETWORK

BOILER

TURBINE HP + MP

ALTERNATOR + HV. TRANSFORMER

C = GENERAL CONTROLLER
BC = BOILER CONTROLLER
TC = TURBINE CONTROLLER
VC = VOLTAGE CONTROLLER
θ_S, θ_R = STEAM TEMPERATURE Π = STEAM PRESSURE
G_W, G_A, G_S, G_F = WATER, AIR, STEAM, FUEL FLOW
n, V = SPEED AND VOLTAGE
P_C = LOAD SET POINT
S_C = SUPERHEATER CONTROLLER
R_C = REHEATER CONTROLLER (NOT SHOWN)

P,Q = ACTIVE AND REACTIVE PRODUCED POWER

FIG. 2

steam temperatures, furnace pression, oxygen content in flue gasses, drum level or steam percentage at outlet of the evaporator (once-trough boilers), ..., and other quantities difficult to measure, as metal temperatures in the boiler. Besides this, other state variable caracterise the turbine set, i.e. speed (or frequency), active and reactive power, voltage, ..., for the short-term aspects of the dynamics, and metal temperatures and expansions regarding long-term dynamics (load changes and start-up problems) /5/, /6/

At the network level, (see fig 3), we have to consider some stability effects, which are very short term phenomenons, and other slower dynamics aspects as the control of frequency, both related to network topology and impedances, and to the rotating inertia of turbine sets and some dynamic loads (mainly big motors). Beyond this, a static problem of security exists, i.e. the control of power flows in the network itself, normally limited by thermal limits of lines, cables and transformers.

The stability question, dynamic, in case of short circuits in the network, or static (long and heavily loaded lines), is function of the load level of the network, voltage profile, phase and frequency at every node, and of the dynamics of voltage regulators and speed governors of the turbine sets. Several state variables have to be considered to describe these phenomenons including as stated above, statoric voltage and phase, internal fluxes, to cope with transient and sub-transients dynamic effects, frequency, ...

The static security problem (overload of lines) is more easy to describe, two state variables i.e. phase and amplitude of voltage, are enough to determine the flows in the network /7/, /8/.

1.3. Till now this important control problem was relatively easy to solve due to the natural non-interacting properties of each part of the whole system, as depicted at fig 1. This is nowadays less true, new production and network techniques being used, as for instance :

- systematic interconnection of networks of neighbours,
- increase in size of base load power plants, both thermal and nuclear,
- use of fast starting peak turbines,
- appearance of big dynamic loads,
- ...

The overall operation of the production-distribution system is therefore rended more intricate, and less static. Coordination organization created with the advent of network interconnection, have seen an increase of their duties, in parallel with an increase of automatic control techniques. Hierarchical control strategies used by these organization are based on a decentralization of some regulating and supervision functions and centralization of other optimization functions.

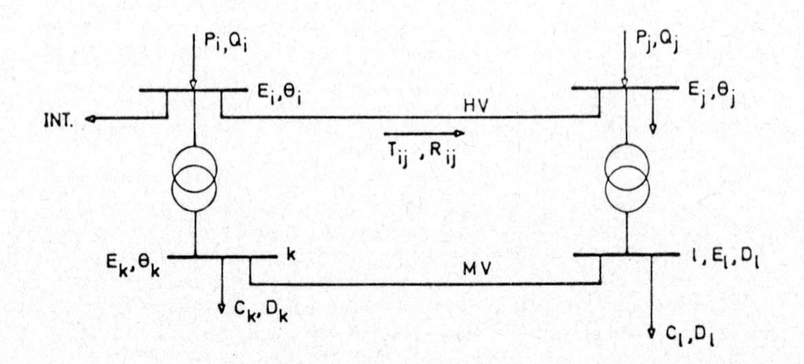

i, j, k, l = BUS NUMBERING
P, Q = PRODUCTION MW, MVAR
C, K = CONSUMPTION MW, MVAR
E, θ = ARGUMENT AND PHASE OF VOLTAGE (kV, rad.)
(STATE VARIABLES)
T, R = TRANSITS MW, MVAR

FIG. 3 STATIC MODEL OF NETWORK.

Let us list some of the features of this hierarchical organization : (Fig 4).

- at the power plant level : (decentralised)
  - all boiler control loops, (pressure, temperature)
  - speed governors and voltage regulators,
  - nuclear reactor control loops
  - automatic start-up and shut down systems for base and peaking units
  - ...

- at the network level : (decentralised)
  - protection systems
  - automatic switchover of loads,
  - transformer voltage regulators,
  - ...

- at the coordination level (centralised)
  - power-frequency control
  - economic dispatch (optimising control)
  - security supervision
  - ...

Let us point out that the coordination level may be splitted in different hierarchical levels itself :

- on a geographical basis, every area of 1000 MW for instance having a local dispatch,

- on a voltage level basis, transport network being controlled by one centralised organisation, and distribution network by an other one.

This is especially true for the static supervision, (security of network and overloads aspects) where there is no closed loop control, but only supervision from one or several dispatches /8/

It is clear at the present day, that the dynamic security of the network (short-term stability as stated above) is essentially decentralised and based on the coordination of protection devices, at the network level and turbosets controllers, at the power station level. It can be foreseen that some backup and adaptive coordination could be centralised in the future /9/, but real short-term network stability centralised control is certainly for the long future, even if some interesting studies are published nowadays on these questions /50/.

1.4. The operation of a power and distribution network is for a control engineer essentially a multivariable problem. Non-interactive control systems, hierarchical control and supervision, are techniques to be used systematically in this field.

2. MATHEMATICAL MODELS - IDENTIFICATION.

The greatest part of the efforts in identification and modelling has been done in the field of thermal and nuclear power plants, /3/, /11/,

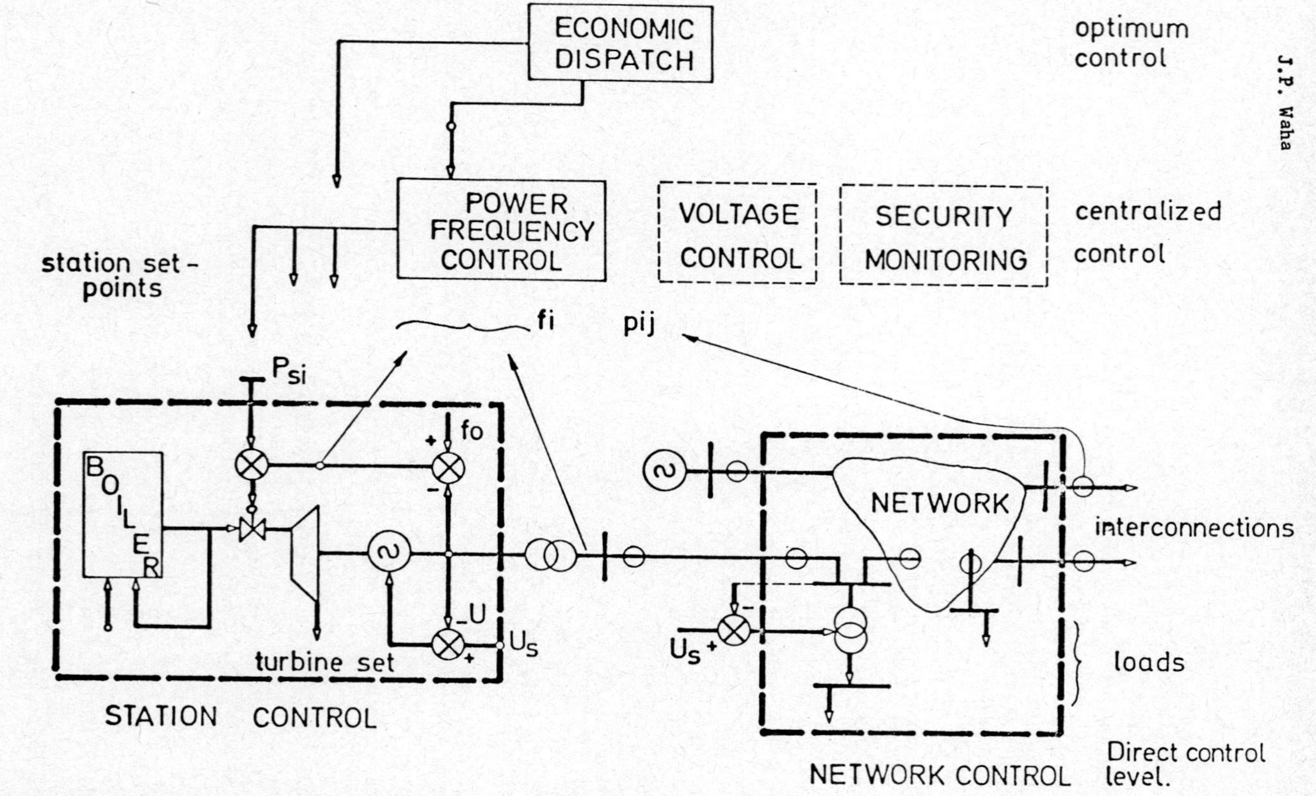

FIG. 4 Multilevel aspects of power system operation

including turbine sets /48/, /49/.

Mathematical models of network, both static for load-flow and security purposes, or dynamic for stability or power-frequency control, where essentially mathematical, and needed no sophisticated identification methods. The reason of this lays in the fact that till now, power-frequency control excluded, all these models were used for off-line studies and planification purposes. But the control of the network, starting with frequency control, and including nowadays on-line supervision techniques indicates that efforts have to be done in this field too, as foreseen in /7/, /25/, /26/, /27/, /28/, /39/, /46/.

2.1. Fig 5 describes a common theoretical model, representative of several industrial situations. The example given is the now classical dynamic behaviour of a steam plant superheater, which is basically multivariable by itself. From the point of view of the linearised behaviour of the influence of input steam temperature on output steam temperature, i.e. the F_1 transfer function of fig 5, classical step response techniques have been widely used, /12/.

This simple situation allows an easy comparison between theoretical and experimental models.

The "multivariable" aspect of the problem rises when one is interested in the two other inputs, steam flow and heat flux, simple step responses being experimentally very difficult to realize. Fig 6 gives a more general model, limited in this case to two inputs, showing the dynamical relationships between inputs, as normally happens in industrial plants; this explains why it is impossible to apply steps response to the x and y inputs of the studied model.

Statistical techniques have therefore been widely used, the aim being at the same time to identify the process during normal operation.

In multivariable cases it is nevertheless difficult to use the normal background perturbations, owing to the high correlation of the inputs. Since then, it has been proposed and successfully tested, to inject some known signal at the input. Different techniques have been used, like the injection of binary perturbations /13/, of non-correlated signals, while working in closed loops /4/, (see B 1 and B 2 on fig 6), or some particular input functions /14/, generalizing the step response technique.

Anyway, the experimental results are obtained after a heavy computing load /15/, /16/, and have to be compared in detail to the theoretical models at our disposition. The analysis of the quality of the results has to be done with great care owing to the experimental difficulties, including bad signal to noise ratio, ill-shaped perturbation spectrum, non-stationarity phenomenon during the observation period, etc.

Systematic multivariable identification of plants, and especially

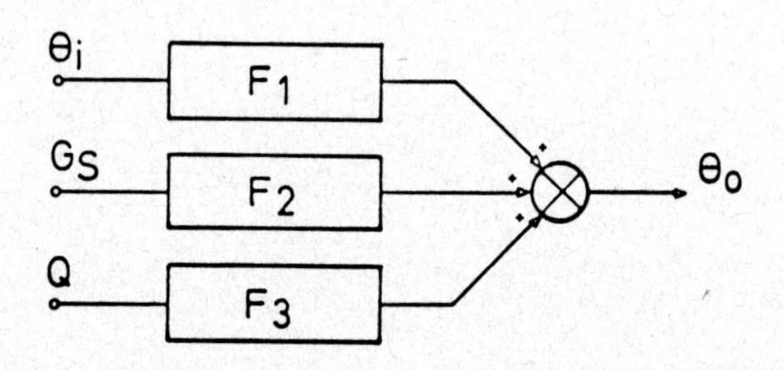

θi (°C) θo (°C)

Gs (kg/s) Q (kcal/s)

θ = STEAM TEMPERATURE i = INPUT, o = OUTPUT
G_s = STEAM FLOW
Q = HEAT FLUX
F_i = LINEARISED TRANSFER FUNCTION

FIG. 5 Superheater model.

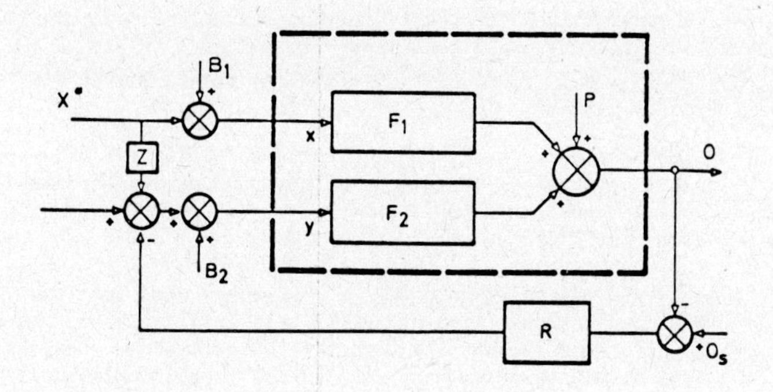

O = OUTPUT

O_s = SET POINT

F_i = TO BE IDENTIFIED TRANSFER FUNCTION $i = 1.2$

Z = UNKNOWN COUPLING } NOT TO BE IDENTIFIED

R = CONTROLER } NOT TO BE IDENTIFIED

P = PERTURBATIONS

B_i = INJECTEDS SIGNALS OF KNOWN STATISTICS $i = 1.2$

IDENTIFICATION :

CONDITIONS : $\phi_{B_1 B_2} = \phi_{B_1 P} = \phi_{B_2 P} = 0$

$\phi_{xy} \neq 0$

TO BE SOLVED SYSTEM :

$$\phi_{B_{1o}} = F_1 \phi_{B_{1x}} + F_2 \phi_{B_{1y}}$$

$$\phi_{B_{2o}} = F_1 \phi_{B_{2x}} + F_2 \phi_{B_{2y}}$$

FIG. 6 Multivariable system identification

those having once-through boilers is far from being terminated /20/; in this special field sensible results have been obtained in theoretical modelling, /10/, /21/, quantities till now neglected being includéd in the models of big units (the influence of pressure on the dynamics of evaporators and superheaters of these units, for instance).

Comparison of statistical identification procedures and deterministic ones have been done for one-dimensional problems /17/, /18/; at the same time the influence of accuracy of identification on control quality has been studied too. /19/.

On the multivariable aspect of the problem, few theoretical results are available; nevertheless comparison between theoretical models and experimental results of complete boiler-turbine systems by digital simulation, has given good results. /55/

2.2. Parameter estimation techniques /16/, nowadays used more and more, accomodates very well with state variable description of the models. They give good results for concentrated parameter processes, the comparison between experimental results and theoretical models being more easy than for the case of distributed processes. Theoretically speaking, there are some difficulties when building a model for parameter estimation, a first hint being the choice of the order of this model. Some interesting results given in /22/ allows to check if the selected order is representative of the dynamics of the system.

In the field of boiler identification experimental results, and comparison with other techniques are given in /17/.

This parameter estimation scheme may be used for adaptive estimation, as shown a few years ago by Kalman /23/. Experimental results are given for the boiler industry in /24/, the examples described being those of drum-level dynamics, and boiler pressure loop control. Notwithstanding these results, there is actually no evidence of systematic industrial use of adaptive identification techniques in the power industry, at least at plant level.

2.3. The operation of networks leads to several identification problems, the distinction between them being the values of time constants involved in the process. We will speak of long term dynamics when dealing with the thermal behaviour of transformers, cables, overhead lines, the time constants varying between a few minutes (5 ... 15 min) for lines, to hours (oil time constant of transformers), and sometimes more for underground cables.

Medium term dynamics will include all aspects related to the rotating energy of the network when dealing with power-frequency control, the time constants being of the order of a few seconds to one minute, seldom more.

Short term dynamics, describing phenomenons in the range of 100 msec or even less, to a few seconds, includes all aspect of network stability, especially in fault cases, heavy interaction between rotating energy, transient behaviour of speed governor and voltage regulator, and protection being involved.

2.31. For short term phenomenons there is few experimental identification work done, mathematical modelling of machines and networks being worked out from constructive datas; this is not always the case for turbine dynamics /48/, /49/, some work being done in the comparison of theoretical models and experimental results of governors, nor for load dynamics.

Load dynamics, including active and reactive power variations with voltage and frequency are needed both for short term and medium term effects as described above. Mathematical models of big loads, like synchronous or asynchronous motors, based on the Park equations are well known, but few things are done to cope with the transient behaviour of composite loads, the problem being statistical in nature. Some experimental results are given in /56/, with respect to static and transient behaviour of loads with regard to voltage changes; a more general theoretical study including frequency effects is given in /57/, USA statistical results are given in /58/.

2.32. Mathematical models of networks, including turbine dynamics, and power-frequency control are described in /28/ and /39/, the multivariable aspects being detailed. It is rapidly seen that the analytical work cannot been carried very far, due to the high dimensionnality of the process, and the quantity of parameters involved. These models give guidelines to the experimental studies which have to be done to overcome the analytical difficulties. Mean values are obtained for network droop and primary regulating energy. The interpretation of the experimental results is difficult, even if the experiment itself is simple. The well known determination method of isolating a network from its neighbours, by opening a loaded line, and recording the frequency variation (see fig 7) neglects a lot of transient phenomenon, interaction between the two pairs of variables : frequency-active power and voltage-reactive power, and non-linearities of speed-governors (dead band) /49/

Theoretical studies coupled with statistical experiments have shown the difficulties of interpretation /26/, /27/

Results obtained from the Belgian network by the classical method gives roughly a value of 400 MW/Hz for the primary regulating energy (datas valid for 1965). Statistical studies of correlation between mean frequency and exchanged power with neighbouring countries give values in the range of 200 to 300 MW/Hz, this being done for several

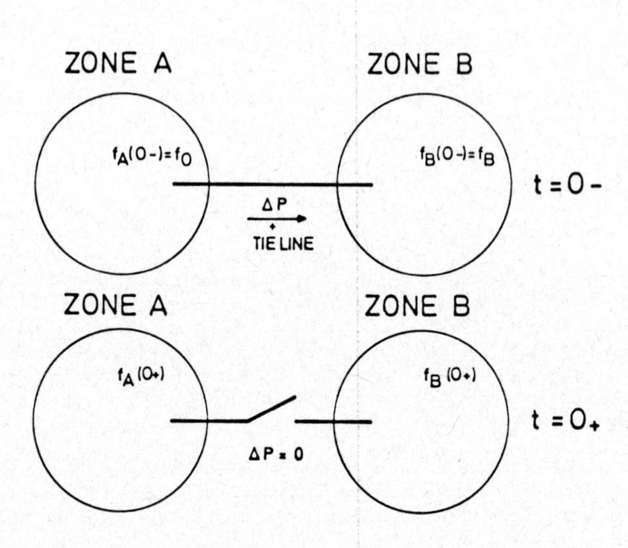

REGULATING ENERGY :

ZONE A: $$E_A = \frac{\Delta P}{f_A(0_+) - f_0}$$

ZONE B: $$E_B = \frac{\Delta P}{f_0 - f_B(0_+)}$$

MW / Hz

FIG. 7 Primary energy deterministic identification.

different days and load conditions /25/. Other results, obtained during holyday period, at low load have given results as low as 100 MW/Hz, and unexpected phase-shifts between exchanged power at different boundaries. Some not well understood dynamical effects show that the basic hypothesis of the experiment, i.e. the one-frequency area, with stiff links between machines is not always true.

The statistical measurements described above were done in conditions where the standard deviation of frequency never exceeded 10 mHz, value well in the range of dead band of speed governors.

Let us point out, that following the well known relationship between network droop and regulating energy E_R.

$$E_R = \frac{1}{s} \frac{Pu}{50}, \text{ (Pu = produced power)}$$

the 400 MW/Hz value gives a network droop of 20 %, when experimental results of turbine droop measurement give values in the range of 4 to 7 % /49/ : this favours too the hypothesis of the effects of governor non-linearities.

It is shown in /27/ that the influence of the reactive energy is not to be neglected, correction formulas being given for the case of the deterministic experimental method, errors of 200 % being quoted if these corrections are not made.

2.33. In the field of long term thermal transients of lines, cables, transformers, there exists too experimental difficulties.

Let us point out for overhead lines the problem of knowing at all time the weather conditions along the line, and for underground cables the well known problem of soil moisture. /53/

Till now, thermal models are used mainly for constructive purposes; integrated control of the network needs simplified models, analog /52/ or mathematical, for on-line computing.

Some results /51/ obtained from experimental datas recorded on an interconnection transformer have shown the possibility of off-line identification; the method used was the quasi-linearization technique, giving a good fit to the parameters of a known theoretical model.

3. AUTOMATIC CONTROL AND SUPERVISION TECHNIQUES FOR MULTIVARIABLE SYSTEMS.

3.1. INTRODUCTION.

Multivariable control techniques have found a wide field of application in power plants, and maybe less in network operation, if we do not include power-frequency control.

Thermal and nuclear power stations, receiving their production set-point from a coordinating organization, inject their power in the network. They have the responsability of their own quality, economicity and security, and are not involved with the network operation. Every increase in automation of the power station units for fixed, variable load operation or start-up and shut down procedures, etc, will contribute to the overall quality of the power and network system; the return of an increase of automation has therefore to be studied not only at the plant level but too at the general one. It is therefore hard to judge at present days if multivariable techniques are sound on an economical basis, or not.

A simple example will clarify this : since a few years all over the world automatic start-up and shut down procedures for power plants have been tried, actually with success, see for instance /1/, /5/, /6/, /29/, /30/.

A basic study made in Europe, /11/, has shown that the "will" of automation was clear, but that the aim was different from one country to the other :

- It is stated sometimes that automatic start-up techniques have to be used whenever the plant has frequent start- and stop cycles; this allows better control, less damage to the material, and uses less operators.

- Other companies think that these techniques have to be used for base load plants, with unfrequent stops, the operators having less experience for these difficult procedures.

Technically speaking, multivariable techniques in the broad sense, i.e. including non-linear and sequential operation, have shown their possibilities, but owing to difficulties in appreciation, it is not yet possible to estimate their returns.

3.2. MULTIVARIABLE CONTROL IN POWER PLANTS.

3.21. The wish to control a power plant as a whole is well known; this leads sometimes to reconsider the choice of the controlled variables.

An early study /31/ has given results on the aims of an integrated control of a plant, and shown some interesting control structures as applied to supercritical once-through boilers. This "Direct Energy Balance" method is widely used nowadays /11/, and has to be compared to other techniques, more familiar to the theoretician of multivariable problems, like the open-loop control system /32/ and some application of non-interacting control to boilers /33/

Primitively, the two last techniques were designed for constant load operation, including linear models with constant parameters; it has been shown that open-loop control leads to non-interaction /33/.

Fig. 8 gives a description of this approach in the case of two dimensional system, using the so-called "model" control techniques /30/.

When one intends to automate load changes, or even start-up procedures, we are faced with two problems :

- the automatic changes of set-points of the plant, which can be worked out from the Direct Energy Balance method, and leads to the so-called "Coordinated Boiler-Turbine control" (fig. 9).

- the normal non-linearities and parameter changes of the system, at different load-level, which can be more easily overcome with an extension of the model technique /30/.

A good knowledge of the dynamics of plants for abnormal load (i.e. steam flow) and pressure conditions, is needed for such automatic procedures. /3/ gives for instance reliable theoretical models for superheater for these cases.

These two approaches will have soon to be integrated, especially for once-through boilers where the natural interaction of different parts of the system are high.

3.22. An interesting example of non-interacting control is given by the control of the furnace of a "non-welded" boiler. The problem is to control at the same time the furnace underpressure, and the excess air ratio, the controlling variables being air flow and flue gasses flow at the exhaust of the boiler. It is felt and experimentally checked that the system is highly interactive and that the dynamics of both paths are of the same order. A control scheme like the one sketched on figure 8 allows good control, with non-interaction, but uses more hardware than classical one-loop feedback systems; the results obtained are interesting, allowing to operate the boiler with a better control of the excess-air. It is worthwhile to note that the non-interactive system, based on "model-theory" is more flexible than others, the operation with one loop on "hand control", and the other one in "automatic control" being possible without difficulties.
An example of coordinated control, which is not non-interactive, is the transient fuel flow-water flow control of once-through boilers /20/, the problem being to adjust water flow to changes both in boiler load and furnace heat flux at the same time. This control problem is

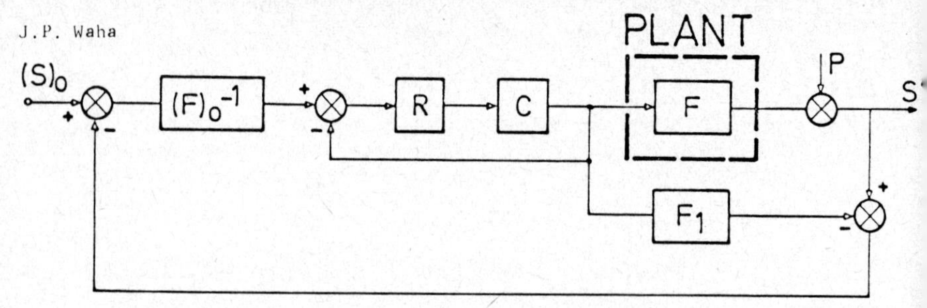

Single-loop model control.

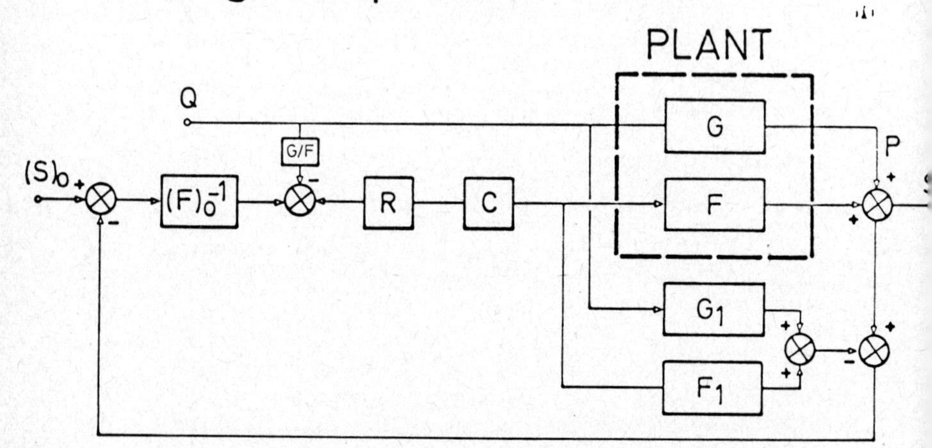

Single-loop control with open-loop influence of measurable perturbation.

F,G = PLANT TRANSFER FUNCTION

F_1, G_1 = MODEL OF PLANT

$(F)_o^{-1}$ = APPROXIMATE INVERSE MODEL

C = ACTUATOR

R = INPUT REGULATOR

FIG. 8a Model approach of feedback control.

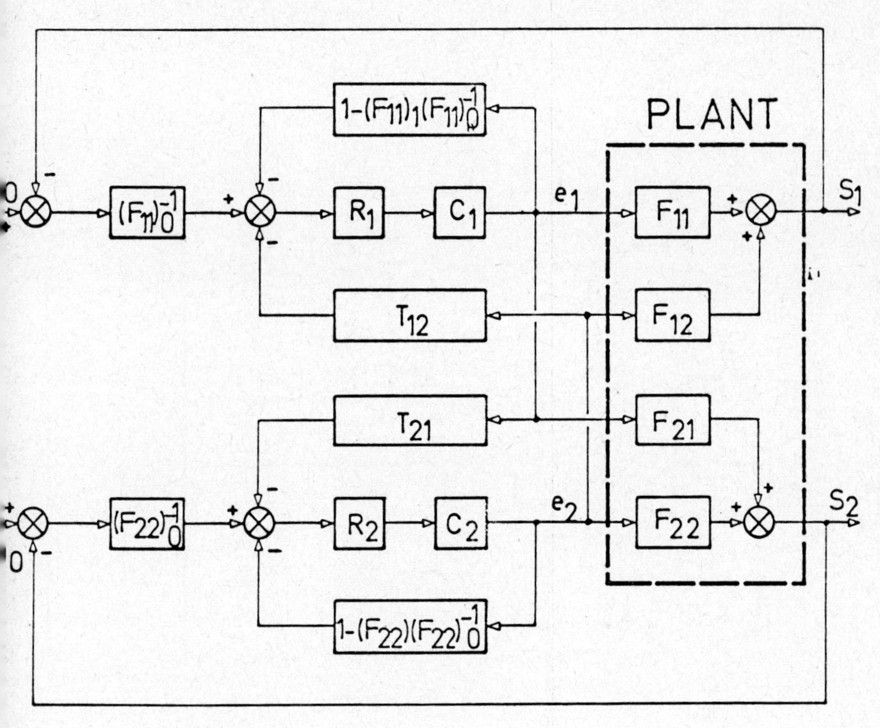

$$T_{12} = \frac{F_{12}}{F_{11}} \left[1-(F_{11})_1.(F_{11})_0^{-1}\right]$$

$$T_{21} = \frac{F_{21}}{F_{22}} \left[1-(F_{22})_1.(F_{22})_0^{-1}\right]$$

S_1 = FURNACE UNDERPRESSURE S_2 = O_2 CONTENT IN FLUE GASSES
e_1 = FLUE GASSES FLOW
e_2 = AIR FLOW
C_i = ACTUATORS AND BLOWERS
R_i = INPUT CONTROLLER
F_{ij} = PLANT TRANSFER FUNCTION
$(F_{ij})_1$ = MODEL
$(F_{ij})_0^{-1}$ = APPROXIMATION OF INVERSE MODEL

FROM REF /33/ DEBELLE

FIG. 8b Non-interactive control scheme (based on model theory)

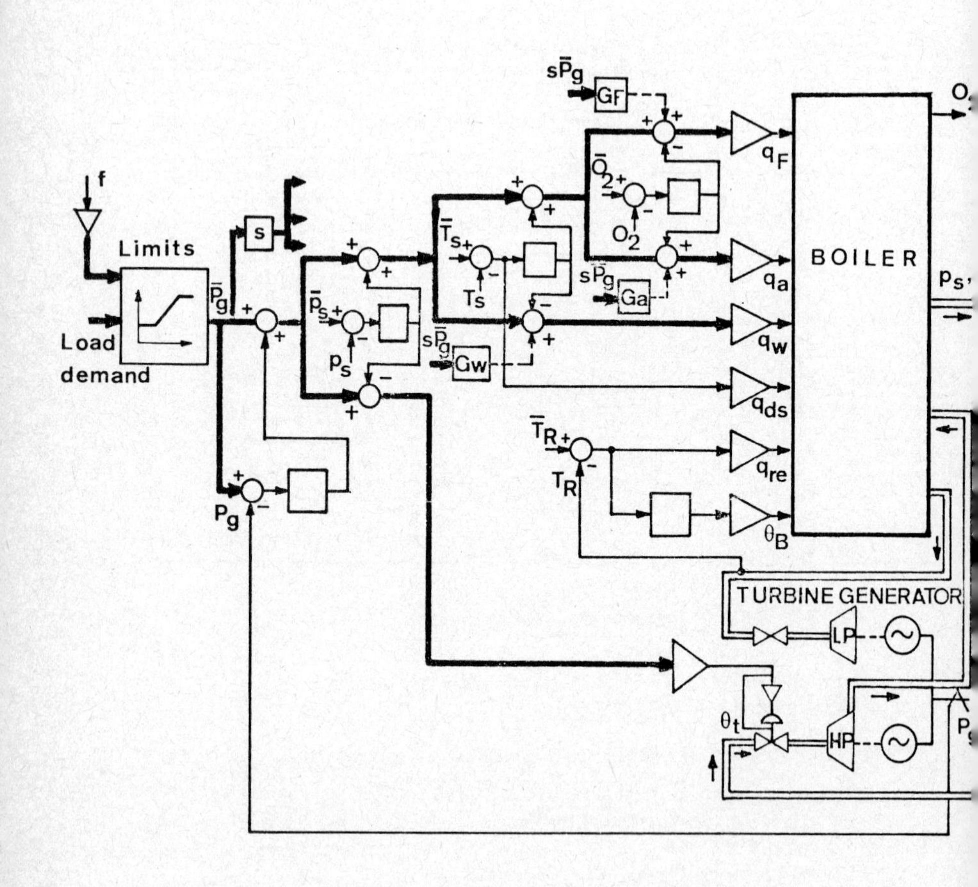

FIG. 9 COORDINATED BOILER-TURBINE CONTROL

FROM REF. 11 - QUAZZA

even more difficult to solve for the new once-through boilers with superimposed water circulation /21/; it is function of the proper states variables to measure, and control. Idealy one would try to keep constant the steam quality at the output of the evaporator, but this leads to difficult measurement methods; it is proposed to control hydraulic quantities (pressure drop on the evaporator, water level in separator for SULZER TYPE Boilers), instead of the classical thermal measurements made actually. Some practical results in the case of a sub-critical unit exists /34/.

It is known that drum type boilers operating at nominal conditions are naturally non-interactive, especially for the steam pressure and temperature loops. This statement does not hold more for great load changes and start-up conditions /30/. Reference /35/, presented at this congress, describes such a case, and shows how it is possible to adapt classical control loops to non-linear and higher dimension systems.

Results are obtained from digital simulation, including the non-linearities of actuators, etc ...

3.23. Start-up and shut down automatic procedures of thermal plants are different for every type of unit, the greatest difference being between once-through and drum-type boilers /5/, /29/, /30/. A complete system would take care of sequential control, structure changes in analog control loops, parameter changes, (gain, time constants) to cope with the start-up procedures. During some steps of the process, it appears that there are sometimes more controlled variables than controlling ones /3/, which is the case when one tries to act on the heat fluxes to control steam and metal temperatures, while rising the steam pressure.

Notwithstanding all these difficulties, there are several units all over the world which have such automatic procedures for their load-changes and start-up.

Nuclear plants, due to economic reasons, are normally running at constant load; full automatic start-up seems therefore not yet to be instrumented. Nevertheless, owing to a better knowledge of the internal behaviour of reactors, theoretical aspects of start-up for given criteria are being studied in detail. Based on detailed mathematical models, it is seen that the start-up problem is described in terms of optimal control, and uses complex computing algorithms. /36/, /37/.

3.24. To end with this chapter, we will note some D.D.C. experiments in power plants. European experiments are described in /11/, a new development being given in /29/. To our knowledge it seems that there is no actual evidence of the systematic use of these techniques for industrial purposes. This is mainly due to price of installation, both hardware and software, boilers having normally few control loops, and to the reliability of a centralised control system.

This could change in the future with the decrease of computers prices, the better reliability of these systems, including dual processor systems, systematic DDC software package available now, and last, some theoretical advantages of DDC control itself (deadbeat control, for example)

3.3. NETWORK MONITORING AND OPERATION.

3.31. Neglecting the optimal control problems, the only control functions are those of frequency and voltage, the first being dependent of the equilibrium between produced, delivered and exchanged active power, the last being function of reactive power.

This problem has to be treated on a centralized basis, including frequency and voltage control at the same time /1/. Practically, the power-frequency control is centralized, and made independent of the voltage control, although the stated interactions between both exist. (see chap. 2).

For the voltage control, besides the normal decentralised techniques of local control by automatic tap changers on transformers, some centralised techniques, called "MVAR" dispatch are running nowadays /1/, /38/.

Frequency control is centralized and combined with interchange control; /39/ describes the multivariable aspects of the power-frequency control of N interconnected areas, regulating energies and start-up times of networks being rigorously defined through a transfer matrix approach.

For the first time the classical "autonomous" criteria of control is compared to non-interactive control between frequency and exchanged power.

There has been some technical evolution of power-frequency control from the very beginning, when analog controllers were used. Digital computers were used in the last fifties, an analog controller working as back-up.

Actually, due to decrease in price and increase of reliability, dual computers systems are installed /45/. The second computer is used as back-up for the first one, and at the same time executes non fondamental work, which can be stopped in case of failure. Besides this, although the digital algorithms used are a transcription of the operation of the analog controller operation, there is easy integration with control of higher hierarchical level, as the economic dispatch.

3.32. The above described control neglects the security aspects of the network operation from the point of view of overload of lines and transformers. Some "automatic" monitoring systems, including security assesment are nowadays running at dispatching level using off-line or on-line computing techniques /41/, /42/, /43/.

Topics of this field includes : /71/

- real-time load-flow algorithms, for process computers,
- redundancy and topological choice of measurements,
- state estimation of the network,
- security assessment by outage simulation,
- etc.

These static state estimation techniques, are based upon Kalman filtering, in a non-linear context, due to the load-flow equations. Some improvements of this static estimation has been done by including the slow transients due to load changes /46/, giving a so-called "dynamic tracking estimator".

Static estimation was based on redundancy of measurement and spatial coherence techniques; the dimension of time included within the dynamic estimator takes care of the natural coherence of the load fluctuations (covariance matrix), besides the spatial coherence, and paradoxally decreases the computing load.

We have to remember that these methods give to the operator a knowledge of the instantaneous security of his network, and through outage simulation of prospected situations. There are in no way control techniques involved, the feed-back being always made via the operator.

The control vector is, by the way, very complex, the operator having at disposal several actions, i.e. :

- production changes,
- load transfer from one high voltage substation till another one,
- topological change in the network,
- in emergency case, load schedding,
- etc.

The technique of outage "simulation" gives time to the operator to take a decision, and eventually to check it by his on-line computing facilities. Full automation, includes first a systematic choice of the control action, and needs therefore an objective function to be optimised (see chap. 4), or may use heuristic techniques /47/.

3.33. SHORT-TERM DYNAMICS. STABILITY PROBLEMS.

Stability of a network is resolved in the 100 msec to 1 sec after a fault has occured, and successfully been eliminated or not. In such a short time, it is impossible to transmit real time data from the substations to the central dispatch, to compute the actual state variables, to run an on-line stability program, and with an "unknown" algorithm to look for the correct action to be executed (line opening, load schedding, etc.).

Therefore, exactly as for the static security assessment, predictive on line outage simulation were proposed /50/. Trials on an IBM 360/95, with a network of 30 nodes, using a simplified stability program based on Park equations, a simulation of 50 outages took more than 15 minutes, which is far too long if we think to the computing time needed by a process computer compared with the 360/95.

Since then, Pattern Recognition techniques have been proposed, based on systematic off-line stability analysis /72/. For a given class of outages, it is possible to build a decision algorithm, function of the steady-state variables, which indicates stability or not.

In practice, the systematic off-line simulations define in the state-space "stable" and "unstable" areas, which are delimited by the "decision" function; finally correlation techniques are used to reduce the state space dimensions to a limited number of important variables.

On line, from transmitted measurements, one would compute the state of the network, as for static security assessment, and with the decision function check whether the system is potentially stable or not. This can be done on a medium process computer, and does not require a heavy computing load. It is to be emphasised that this technique gives no guideline to the operator as how he has to control his network to go from a potentially unstable to a stable state.

3.34. Network control includes nowadays security monitoring, power-frequency control and on-line security assessment, being treated at the same time, at the dispatch level.

It is worthwhile to note that besides complicated mathematical techniques, a lot of data presentation techniques, taking human engineering aspects in account, are used.

Reference /69/ gives for instance a standard list of datas to be transmitted, treated and presented to operators. These datas include substations, power stations quantities, and general network informations (time error, wheather forecasts and actual values).

Operators being unable to look at those quantities at the same time, appropriate scanning rates and presentation have to be selected /71/

4. OPTIMAL CONTROL OF THE OVERALL OPERATION OF PRODUCTION, TRANSPORT AND DISTRIBUTION OF ELECTRIC POWER.

4.1. Each part of the considered process being controlled with local criteria of security and sub-optimality, the overall optimal control will be performed at the top of the hierarchical coordination scheme. Economic dispatch was for instance, the first example of developed optimal control strategies.

It is worthwhile to study this multivariable and structured process of production, transport and distribution as regards to the different states of operation.

References /59/, /69/ and /70/ define in this perspective the levels of operation :
- direct,
- optimal and
- adaptive

and the states of operation :
- preventive
- emergency
- restorative

Table 1, adapted from reference /70/ sketches the relations between levels and states as defined above, and as they are used nowadays (see page 190).

Direct control has been studied in a classical way in chapter 3 of this survey; optimal control strategies will be the center of this one. Adaptive control works out every modification of the two other control levels, and is strongly related to the identification techniques described in chapter 2.

The hierarchical concepts given in this paper are based on direct control at the lower level of the system (plants, substations), and coordination and optimization at the higher level.

Optimal control, related to the definition of cost functions, minimizes production cost, transports costs (losses), ..., for given quality and security of the distributed power.

It is hard to define costs functions for these two last parameters, as every planification engineer knows. Constraints on state variables are therefore used, in conjugation with cost functions.

4.2. OPTIMAL CONTROL IN THE PREVENTIVE STATE.

4.21. At the power station level great care is given to the operating conditions (steam temperature and pressure) to keep working the boiler-turbine sets at high efficiency, direct control being systematically used in this optics.

Few things have been done to measure the efficiency in transient

Table 1 MULTI-LEVEL CONTROL OF POWER SYSTEMS

LEVELS / STATES	DIRECT CONTROL	OPTIMAL CONTROL	ADAPTIVE CONTROL
Preventive state	Load-frequency control speed governors, generator voltage regulators Plants control loops and automatic sequences ---	- Power plant yield optimization - Unit commitment - Economic dispatch - Voltage control and Mvar dispatch ---	- Set point adaptation - Controller and Relay adjustments - Off-line studies - Manual adaptation
Emergency state	Fault clearing by protection devices (relays, breakers) Load shedding Generator shedding System splitting ---	Maximum load solution	-
Restorative state	Load transfer Peak units start-up Feeder restoration Plant resynchronizing ---	Optimal Restoration procedure	-

conditions, like load changes and fuel changes.

Transient efficiency and start-up costs are nevertheless very important parameters in the overall optimal control of plant and network, the latter being studied with more detail.

Reference /60/ gives results of experimental work from the C.E.G.B., the aim being to optimize the combustion of a boiler-turbine set. The strategy looked for, was the ideal transient excess-air ratio needed to increase the overall efficiency, the calorific content of the burned coal fluctuating much.

Correlation and hill-climbing techniques were used; practically it was established that the method would give savings only if the coal was provided from different mines, with great variations in heat content.

4.22. At the coordinating level, optimal control is dynamic in nature, even if the generally used methods are static.

Since the first "Economic Dispatch" centers, which scheduled the production of plants on order of merit bases, several new problems have been included to the optimal control scheme :

- the reduction of transport losses, using the "B" matrix method, and nowadays optimal load flow techniques /63/, /72/.
- reactive power dispatch and voltage control /1/, /38/.
- security of transport /65/, /69/, /71/.
- load fluctuations with respect to plant dynamics /61/.
- mixed hydroelectric and thermal stations dispatch
- start-up costs and unit commitment
- ...

There is progressively integration of static optimization techniques with dynamic control. If we could assess the cost of load changes of plants, the problem of economic dispatch and power frequency control would merge in one dynamical optimal strategy. Some simplified algorithms have been proposed /61/, but the choice of formulas giving transient efficiency is still difficult. No practical implementation is known.

If we want to limit the problem to a static one, or to a succession of static states discarding all transients, it appears that the search of its solution is stated formally as a non-linear programming problem, including various constraints on controlling variables (power production, extreme position of tap changers, ...) and states variables (power flows on lines and transformers, -).
At the same time there is a mixing between continuous, but limited, controlling variables, and integer ones (stop and start-up of units, topological changes of network, ...)

Proposed computing techniques are of different types :

- heuristic and non optimal, which may solve the problem of security assessment, stated in chapter 3. /47/
- iterative techniques, mainly based on gradient techniques /1/, /72/, which are used to solve the general problem
- linear programming techniques used to solve some linearized approach of the general problem /64/, /65/, /71/
- special techniques, like the branch and bound algorithms, suited to mixed variable problems - /73/

Some of these techniques, using dual variables, give as return the incremental cost of constraints, and are therefore valuable tools for the planning engineer too.

Reference /72/ describes the general problem of constrained optimal control of real and reactive power on a static base. Optimal load flow is the center of the proposed method, using Newton-Raphson algorithms and sparsity programming techniques. Controlling variables are slack bus voltage, active power generation, phase-shift transformers, tap changers, ... State estimation techniques as proposed in / / are not included; hard constraints on state variables are replaced by penalty functions (soft constraints). The iterative loop includes the load-flow as stated above, Lagrange parameter and gradient computations, and finally control variable corrections.

Reference /64/ describes a method of economic dispatch, including power flow and production constraints, the network equations being linearized and treated by the DC load flow approximation. Cost functions are supposed to be linear, but estimates of losses based on quadratic formulas are added to node consumption. The algorithm uses a succession of linear programs, the first giving a feasible solution which is then improved by including the losses.

Computing time of 20 s on CDC 6600 for 100 nodes and 300 lines are given; this method is therefore manageable for planning purposes, and maybe not for on-line control.

A more general approach given in /66/, presented at this congress given an optimizing scheme for hydroelectrical and thermal power systems, energy constraints on the water consumption being added to the problem.

Production costs are represented by quadratic functions of produced power; the network flow limitations are linearized, losses are included in the optimization.

The optimization is not purely static, but merely includes a succession of static stages, to cope with the energy constraint on a period of 24 h. (Discrete maximum principle). The used algorithm splits the optimization in active power optimization, produced power being the controlling variable, and reactive power optimization, active power losses having to be brought to a minimum. A final step of the solution consists in an improvement of the real power optimization with regards to the founded active losses. This approach should be compared to the previous one in computing power and time needed.

An interesting development of DC load flow techniques using linear programming methods are used to solve the more general problem of security including forced outages /71/, /65/.

The second reference combines economic dispatch, running-spare-capacity, with security of intact network and a class of outages conditions. Fig. 10 gives the flow diagram of the procedure.

Due to size of equations and number of variables, especially with outages conditions the dual problem is solved, using the revised simplex method.
The reference gives a computing time of 11,5 s on an Atlas computer, for a problem of 23 nodes, 24 lines and 24 generators, all single outages being considered.

Reference /73/ describes the more general problem including unit commitment of a thermal and hydraulic power system. The hypothesis are :

- cost functions of thermal plants are linear with production
- start-up costs are exponential with shut-down period
- energy constraints on daily water consumption
- DC power flow approximations are valid
- running-spare-capacity are used as constraints

This paper gives an interesting study of various cases, basically of 10 nodes, 13 lines, 24 thermal sets and 4 hydraulic turbines; the branch and bound algorithm is used to cope with this mixed-variable problem. Unhappily no computing duration was available.

This does not cover all the subject but gives some indications on the state of the art of these optimizing techniques, a more complete set of earlier references being given in /1/.

4.3. OPTIMAL CONTROL IN EMERGENCY AND RESTORATIVE STATES.

When a power system is in emergency, or after a great disturbance has to be restored, time is the main parameter to optimize : all control efforts have to be exerced to bring the system back to preventive state in a time as short as possible.

For the "Emergency" state it is nevertheless possible to search for a "Maximum Load Solution" (see Table 1 and ref. /70/), function of disponible production, interconnection interchanges, while the load is satisfied at maximum, avoiding line or transformer overload and bad voltage profile. The objective function to be maximized is non linear (sum of load MVA), like most of the constraints (production

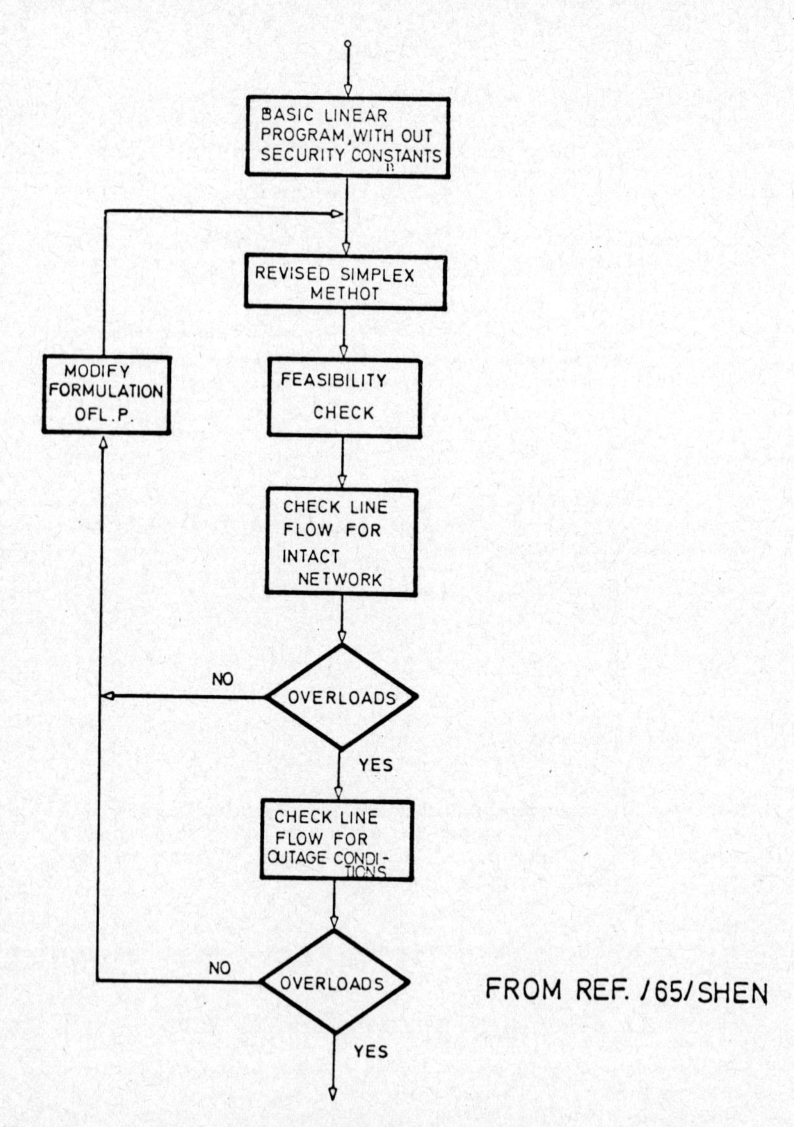

FIG.10 LINEAR PROGRAMMING APPROACH OF LOAD WITH SECURITY

and voltage limits, flow limits, ...). The authors have used the Fletcher-Powell gradient technique, and replaced the limits by soft constraints in form of penalty functions.

Other approaches of the same problem are based on optimal load shedding algorithms /67/, the cost function used being to our opinion, delicate to formulate.

For the basic restorative procedures heuristic methods looking for realisable (i.e. not overloaded) paths through the network can be used /47/; one could imagine that optimizing schemes could be utilised by defining priority amongst the different loads.

We should note that emergency and restorative techniques are by essence on-line, when preventive ones can be used off-line with predicted datas.

We therefore think that optimal control will develop itself in the field of preventive control, although using iterative procedures which are time consuming. On the emergency and restorative side, due to size and time limitations, we think that heuristic methods giving realisable solutions will have to be used in the future.

5. CONCLUSIONS.

There exists strong relationship between automatic control methods and planning ones, chapter 4 on optimal techniques has shown the similarity of the problems.

In the overall economic optimization of power systems it is worthwhile to remember that planning studies are based on availability figures for plants, and recently for the network itself. Every increase in automation giving as result an increase of availability will decrease in the long term the needed investments. Some new definitions of "risk indices" given in reference /74/ are essentially based on this approach, including dynamic behaviour of networks, plants and controlling devices. This will be the way in the near future to answer the question of chapter one on the evaluation of economic return of automatic control.

Finally, it is to be remembered that if power systems are planned on bases of economics, security and quality of energy, using sophisticated optimal search methods, they have to be operated with the same criteria, otherwise the overall benefit of planning and operation of the system would not be obtained.

REFERENCES.

/1/ QUAZZA G. The state of the art in automatic controls for Electric Power Systems - Survey Paper - 4th congress of IFAC - Warsaw - 6/69

/2/ J.T. TOU Modern Control Theory. Mc Graw - Hill 1964

/3/ DEBELLE J. et WAHA J.P.
L'analyse des processus et son application à la détermination des fonctions de transfert des chaudières - Facteurs déterminant l'aptitude à une bonne régulation. Journées AIM - Centrales Thermiques - 11-15/5/70 - Liège (Belgique).

/4/ JACQUET B. Problèmes posés par la détermination de la densité spectrale - Annexe 2 à /3/.

/5/ BERLEMONT V., LAIRE C.
Expérience acquise dans le démarrage automatique des groupes thermiques modernes.
IBID.

/6/ BERLEMONT V., DEBELLE J.
Some aspects of the automatic control of the start up of a drum type boiler. IFAC SYMP. on Multivariable Control Systems.
Oct. 68 - DÜSSELDORF.

/7/ PESCHON J. et al.
State Estimation in Power Systems
Part I and II
IEEE Summer Power Meeting
Dallas - USA - 6/69

/8/ VAN MIEGROET P.
Projet DESCARTES - Apport des installations de calcul à la conduite des réseaux.
LABORELEC - Belgique - Note d'information n° 18 (à paraître).

/9/ UNGRAD H., GLAVITSCH H.
Protection de réserve et surveillance de la sécurité de réseaux coordonnés et centralisés comme constituants pour l'automatisation des transports d'énergie.
CIGRE 1970 - rapport 34-03

/10/ F. LÄUBLI, G. ALLARD, D. LE FEBVE
Régimes transitoires dans les chaudières - Calcul des fonctions de transfert et utilisation des calculateurs par le constructeur - Journées AIM - 5/1970 - Liège - Belgique

/11/ H. BALERIAUX et al.
Experience acquired on the automation of large power stations, and particularly automatic start-up and the automatic control of transient conditions.
UNIPEDE - Cannes - September 1970 - Paper II.2.

/12/ WAHA J.P. Méthodologie d'analyse expérimentale des processus industriels.
Journée IBRA - 6/68 - Projet CHANCE - Bruxelles

/13/ BOARDMANN K.D.
The measurement of the dynamic behaviour of a superheater by random signal testing.
in : Automatic control in production and distribution of electrical power.
IBRA - Brussels - 4/1966

/14/ MENAHEM M. Process dynamic identification by the multistep method.
4th IFAC Congress Warsaw 1969

/15/ GODIN P., DAVOUST G.
Identification par entrées-sorties de systèmes linéaires multivariables.
IFAC SYMP. on Multivariable Control Systems - Oct. 1968 DÜSSELDORF

/16/ ÅSTROM K.J., EYKOFF P.,
"System Identification"
2nd IFAC SYMP. on Identification and Process Parameter Estimation - June 1970 - Prague

/17/ GUSTAVSSON J.
Comparison of different methods for identification of linear models for industrial processes - IBID

/18/ WELFONDER E.
Vergleich deterministicher und statisticher Verfahren zur Systemanalyse gestörter industrielle Regelstrecker
BWK, 22 (70) 6 Juni
p. 288 - 297

/19/ ISERMAN R. Required accuracy of linear time invariant mathematical model of controlled elements. 2nd IFAC SYMP. on Identification and Process Parameter - 6/70 - Prague

/20/ HENRIETTE A.
Etude théorique et expérimentale de la dynamique d'une chaudière à circulation forcée.
Journée IBRA - 6/68 - Projet CHANCE - Bruxelles

/21/ VARCOP Some aspects on the control of the subcritical once-through boiler with surimposed circulation.
2nd IFAC SYMP on Multivariable Control Systems - Oct. 71 - DÜSSELDORF

/22/ MEHRA R.K. On-line identification of linear dynamic systems with application to Kalman filtering.
Private communication - Wolf Managment Service - 260, Sheridan Ave - PALO ALTO California

/23/ KALMAN Computer controlled adaptive control systems - in : "Adaptive control systems -
Mac Graw Hill 1,961
Editor : E. MISHKIN et al.

/24/ MIZUTANI H. et al.
On-line real time identification of parameters for thermal power plant.
2nd IFAC Symposium on Identification
6/1970 - Prague

/25/ WAHA J.P. Etude des fluctuations de puissance sur les interconnexions avec l'étranger.
Rapports internes 2/66/4490 et 2/4636
Laborelec - Belgique

/26/ VITEK V. Statistical methods in the automation of the operation control in Power system.
3d IFAC Congress - London

/27/ CUENOD M., QUAZZA G.
Dynamic Statistical analysis of electric power system controls.
CIGRE Working Group A 1 report - Presented in Sydney, 8/68
IFAC SYMP. on System Dynamics and Automatic Control in Basic Industries.

/28/ DONATI F., FIORIO G.
Sur la dynamique des réseaux d'énergie
Vol. XXXVIII - n° 946
Instituto Electrotecnico nazionale Galileo Ferraris
Italie

/29/ MONTAGNA A., BONGHI M.
Computer applications to steam station control in Italy.
Colloque Informatique - UNIPEDE
6/71 - Lisbonne

/30/ DEBELLE J. Réglage automatique des générateurs de vapeur.
Journées IBRA - 6/68 - Projet CHANCE - Bruxelles

/31/ ARGERSINGER J. et al.
The development of an advanced control system for supercritical units.
Paper CP - 63-1409 - National Power Conference - CINCINNATI
Sept. 1963

/32/ CHAUSSARD R.
Utilisation des méthodes d'analyse indicielle et du critère de NASLIN dans le cas des centrales thermiques.
AUTOMATISME - Tome 9 - Sept. 1964

/33/ DEBELLE J. Réglage non-interactif
Projet CHANCE - note interne - EBES 9/69

/34/ HENRIETTE A.
Réalisation pratique d'une boucle de réglage d'alimentation - Projet CHANCE - Rapport interne INTERCOM
Monceau 1970

/35/ SCHULZ R. The drum-waterlevel in the multivariable control systems of a steam generator.
2nd IFAC SYMP. on Multivariable Control Systems - DÜSSELDORF - Oct.

/36/ FOUREAU A.E.
Etude d'un programme non-linéaire réalisant le démarrage automatique optimal d'un réacteur nucléaire à bas niveau de puissance - Revue A IV, 4, 1962

/37/ HASSAN M.A. et al.
Computational solution of the nuclear reactor minimum time start-up problem with state space constraints.
2nd IFAC SYMP. on Multivariable Control Systems - DÜSSELDORF - Oct. 1971

/38/ BARA J. et al.
On-line control of voltage and reactive power flow in Electric Power systems.
Paper 40-1. 4th IFAC Congress - 6/69
Warsaw

/39/ QUAZZA G. Non-interacting Controls of Interconnected Electrical Power systems.
IEEE Trans - Pas 85 - 7/1966

/40/ MYGIND P. Planning and Implementation of a large Load Dispatch System.
Colloque Informatique UNIPEDE - Lisbonne - 6/71

/41/ MORAN F. An experiment in the automatic control of power generation in a limited aera of the CEGB.
4th IFAC Congress
6/69 - Warsaw

/42/ SAMINADEN V., POUGET J.
Application des systèmes d'acquisition et de traitement de l'information à la conduite et à l'exploitation des réseaux d'énergie électrique.
Colloque Informatique UNIPEDE
Lisbonne - 6/71

/43/ PATTERSON W.J.
P.J.M. Operating criteria for system reliability
Pennsylvania Electric Ass., Systems Operation Committee
Hagerstown, Mar. Oct. 1969

/44/ ARRIATI F. On-line monitoring and state calculation of power systems for optimization and reliability of service.
Colloque Informatique UNIPEDE - Lisbonne
6/71

/45/ DOPAZO J.F. et al.
State calculation of Power systems from Line-flow measurements - IEEE Trans - Pas 89 7 Oct. 70

/46/ DEBS A.S. et al.
A dynamic estimator for tracking the state of a power system.
IBID

/47/ AUGE J. et al.
Contribution à l'estimation de la sécurité des réseaux.
Bulletin de la Direction des Etudes et Recherches
EDF - Série B - 1/1967

/48/ DINELEY J.L.
Power-system governor simulation
Proc. IEE, Vol. 1, 1, 1964

/49/ VAN DE MEULEBROEKE F.
Régulations de vitesse des turbines à vapeur.
LABORELEC - Belgique - Note d'information n 17 - 1971

/50/ TUEL Communication at Purdue University - Symposium :
Electric Power in the 1970's May 1970

/51/ Rapport LRP - ACEC - Charleroi
Estimation des paramètres d'un modèle thermique de transformateur - Projet DESCARTES - LR 60 302/1300

/52/ R. RENCHON Protections et Automatismes valorisant les capacités de surcharge des lignes et transformateurs - CIGRE 1970 - 34-02

/53/ B. GEERAERT
Aperçu de l'état actuel des méthodes de calcul de l'échauffement des câbles souterrains à basse, moyenne et haute tension. LABORELEC - 6.1.69

/54/ N. GERMAY, C. VASTRADE
Etude des performances des câbles souterrains.
Synthèse d'études Françaises, Hollandaises et Anglaises. LABORELEC - 2/69

/55/ GODIN et al.
Simulation numérique d'un générateur de vapeur.
Journées Internationales d'Etude des Centrales Electriques Modernes - AIM - Liège - 11-15/5/70

/56/ WAHA J.P. Analyse des coefficients de sensibilité des puissances active et réactive à la tension. Rapport LABORELEC - 2/4705 - 3/68

/57/ LAIBLE Th. Abhängigkeit der Wirk- und Blindleistungaufnahme passiver Netze von Spannungs- und Frequenzschwankungen Bull. ASE, 59, (1968), 2 - 20/1

/58/ The effect of frequency and voltage on power system load. IEEE - Rapport IEEE, 31 CP 66-64-1966

/59/ T.E. DYLIACCO
The adaptive reliability system.
IEEE Trans. Pas 86 - 5/67

/60/ MORAN et al.
Development and application of self-optimising control to coal-fired steam generating plants
Proc. IEE, Vol. 115, 2/68

/61/ CUENOD M. et al.
Optimum Fitting Method and its Application to Dynamic Economic Dispatching of Power Systems.
3d IFAC Congress - London

/62/ CUENOD M. et al.
Adaptive control of Interconnected Power Systems.
4th IFAC Congress, Warsaw
6/69

/63/ PESCHON J. Optimum load-flow for systems with area interchange controls.
Wolf Managment Services. PALO ALTO, USA.

/64/ DODU J.C., MERLIN A.
Méthode de résolution du dispatching économique dans l'approximation du courant continu.
Bull - Direction Etudes et Recherches EDF, Série B, 3/70

/65/ SHEN C.M. et al.
Power system load scheduling with security constraints, using dual linear programming.
Proc. IEE, Vol. 117, 11/70

/66/ BÜHLER H. et al.
Short-term optimization of the operation of hydroelectric and thermal power stations according to the discrete maximum principle.
2nd IFAC Symposium on Multivariable Control Systems
Oct. 71. Düsseldorf

/67/ HADJU P.H. et al.
Optimum load-shedding policy for power systems.
IEEE Trans. Pas 87, 3/68

/68/ NOFERI et al.
La programmation des unités de production pour la couverture des pointes dans les réseaux électriques: paramètres importants et leur influence relative.
CIGRE - Session 1970 - 32-06

/69/ KIRCHMAYER L.K. et al.
L'automatisme, Facteur d'accroissement de la sécurité de fonctionnement des réseaux. CIGRE - Session 1970 - 32-10

/70/ T.E. DYLIACCO et al.
Multi-level approach to the control of interconnected systems.
2nd Int. IFAC/IFIP - CONF. MENTON - 6/67

/71/ Power System Security Monitoring - WMS Project U 931
12/69
Wolf Managment Services - PALO ALTO

/72/ H.W. DOMMEL
Constrained Optimal Control of Real and Reactive Power Dispatch.
SYMP. on "Optimal Power-System Operation" - Sept. 1969
Univ. of Manchester

/73/ A. MERLIN et al.
Méthode de répartition journalière d'un ensemble de moyens de production thermique hydraulique -
Symp. C.E.E. Varna - 5/70

/74/ NOFERI et al.
Quantitative Evaluation of Power System Reliability in Planning Studies.
IEEE Winter Power Meeting
1/2 - 1971. New-York
Paper 71 TP 89 - PWR

A.G.J. MacFarlane

List of ERRATA for and supplements to

A.G.J. MacFarlane
LINEAR MULTIVARIABLE FEEDBACK THEORY: A SURVEY

p.11, in equation (2-15) read -Y instead of Y ,

p.12, in equation (2-17) read -X instead of X ,

p.12, in equation (2-18) read 0 instead of I_r (twice) ,

p.12, in equation (2-19) read -Y instead of Y and
-X instead of X ,

p.15, in the large matrix in the middle of the page
read $-U_2$ instead of U_2 and
$-U_1$ instead of U_1 ,

p.23, equation (3-10) read

$$P(s) = \left(\begin{array}{cc|c} F & U & 0 \\ -HV & I+HW & I \\ \hline V & W & 0 \end{array}\right) ,$$

p.24, equation (3-12) read

$$\det \begin{pmatrix} F & U \\ -HV & I+HW \end{pmatrix} = \det(F)\det(I+HW+HVF^{-1}U)$$
$$= \det(F)\det\{I+H(W+VF^{-1}U)\}$$
$$= \det(F)\det(I+HQ) .$$

Figures 2-7, 2-9, 2-10, 2-11, 2-12, 2-15 and 3-7 should be as shown on the enclosed figures.

Figure 8-1 is missing and is as enclosed.

The following references should be added:

WALTER, O.H.D., 1970. "Analytical and computational aspects of the linear optimal control problem", Ph.D. Thesis, University of Manchester.

WALTER, O.H.D., 1970. "Eigenvector scaling in a solution of the matrix Riccati equation", IEEE Transactions on Auto. Control, Vol. AC-15, pp. 486-487.

WALTER, O.H.D., 1971. "Formulas for linear optimal control over a finite or infinite interval", Electronics Letters, Vol. 6, No. 19.

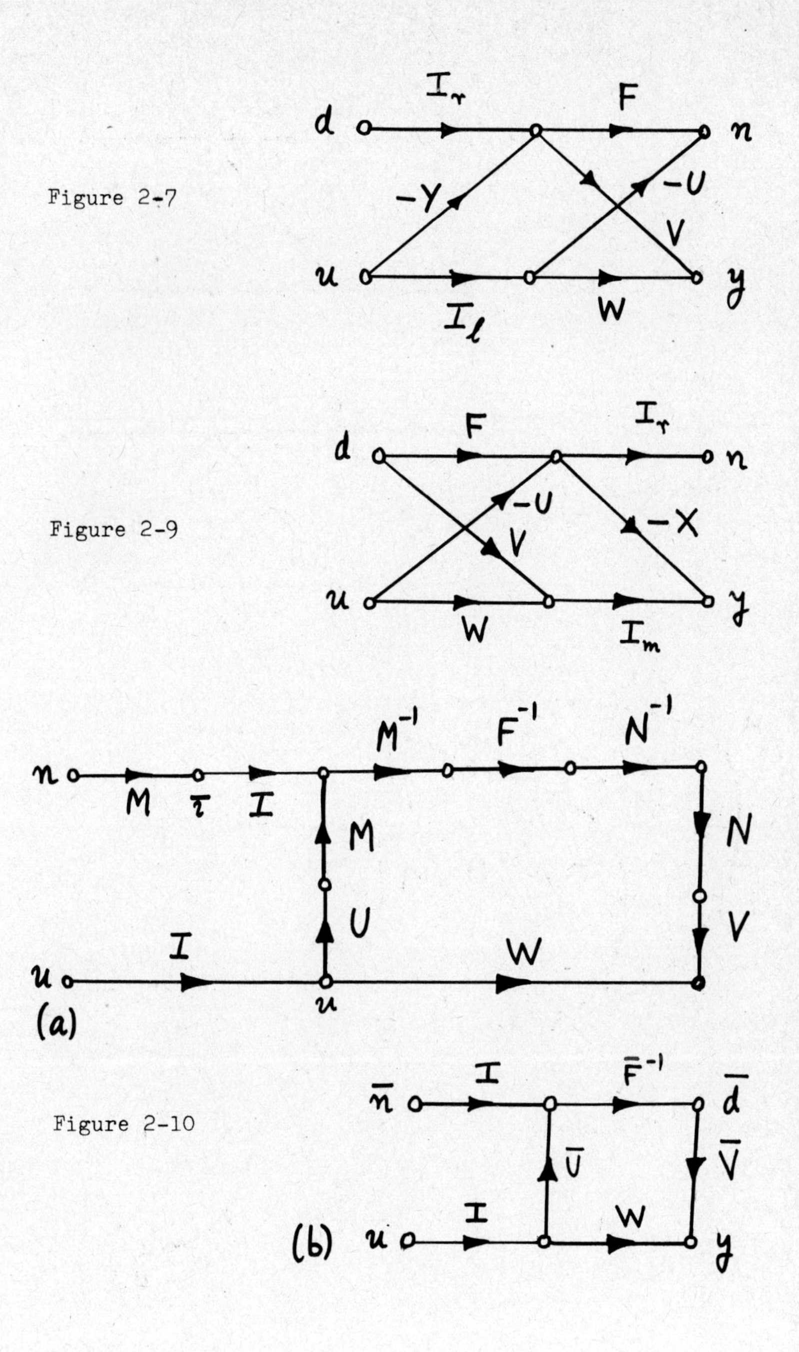

Figure 2-7

Figure 2-9

Figure 2-10

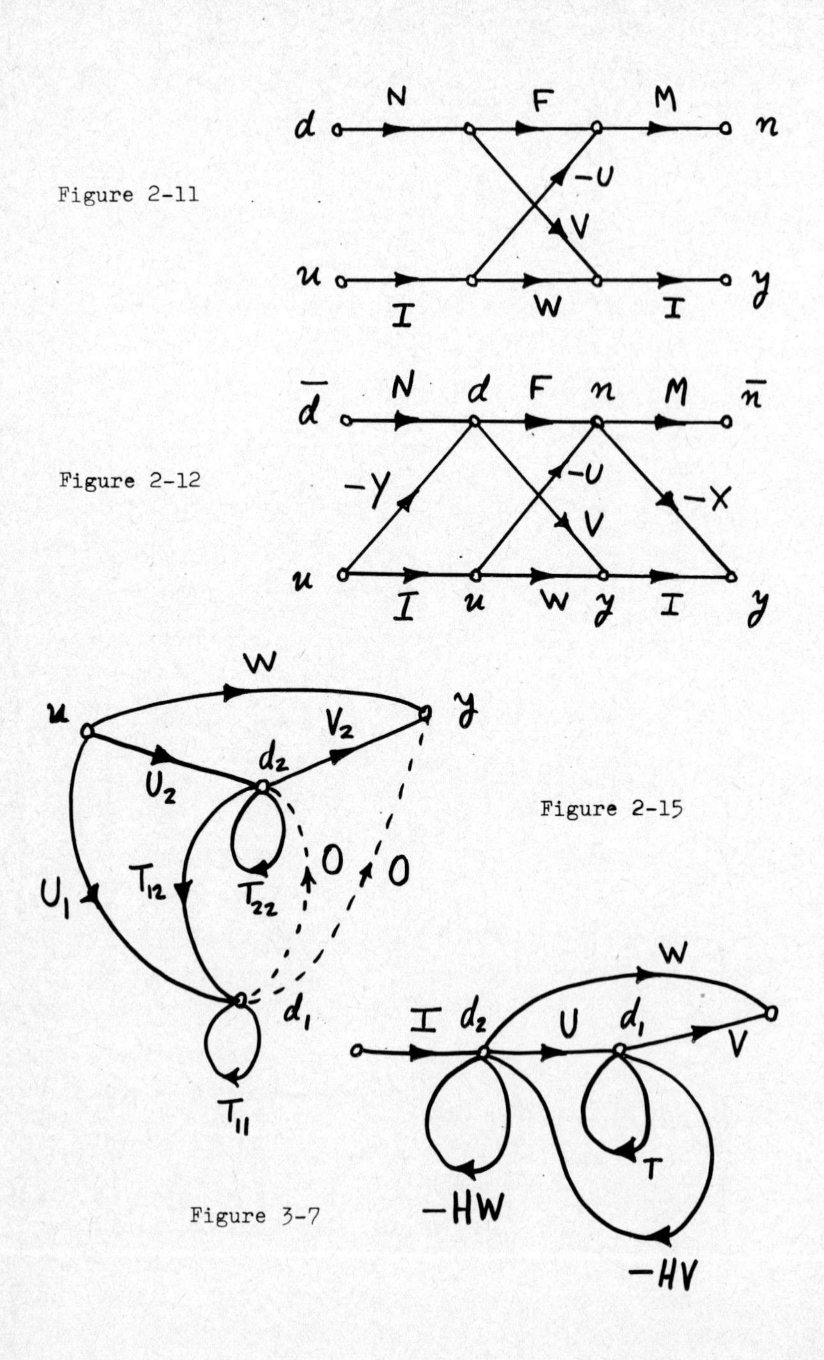

Figure 2-11

Figure 2-12

Figure 2-15

Figure 3-7

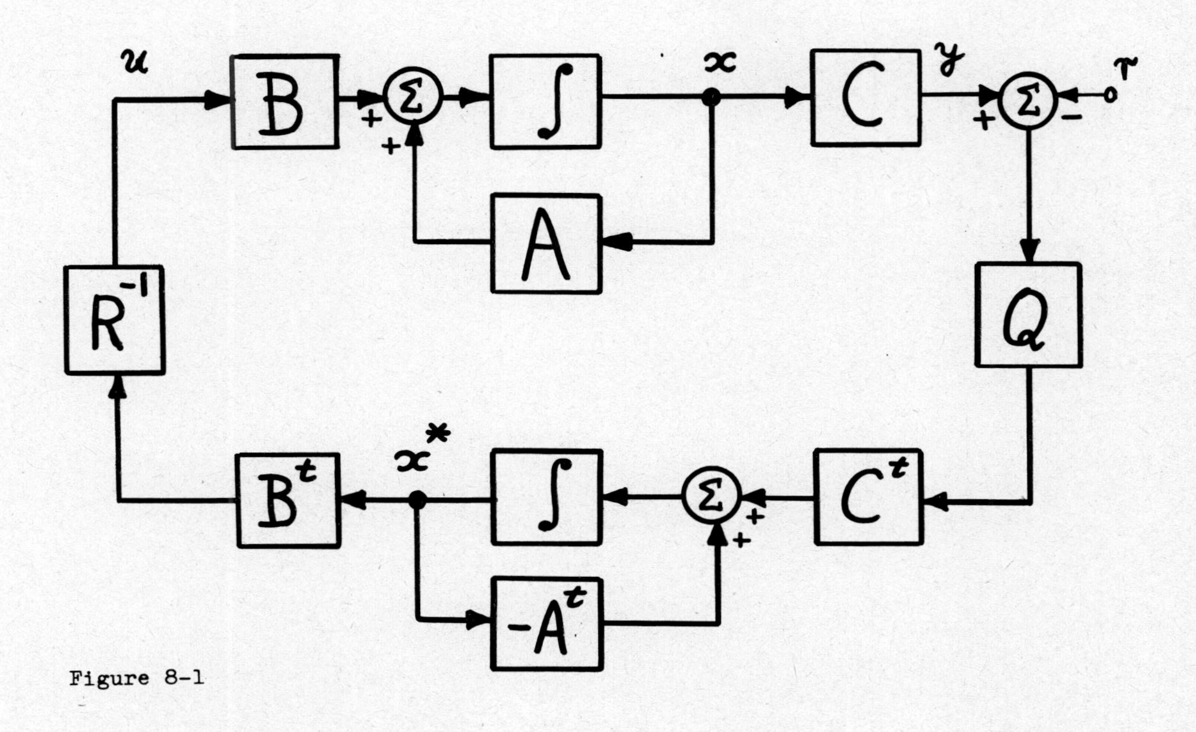

Figure 8-1

M. Thoma

List of ERRATA for

M. Thoma
OPTIMAL MULTIVARIABLE CONTROL SYSTEMS THEORY: A SURVEY

p.83, in equation (2.5) read
non-negativity instead of nonegativity ,

p.83, 1st line after equation (2.5), read
non-negative instead of nonegative ,

p.85, 2nd line after equation (2.12), read
.... which is a global condition .... ,

p.85, 4th line after equation (2.12), read
If instead of Of ,

p.91, 5th line from above, read
equation instead of equations ,

p.91, 11th line from below, read
reachable instead of realable ,

p.96, in equation (3.41) add = 0 ,

p.96, 5th line from below, read
provides instead of proves ,

p.98, 3rd paragraph, 6th line from above, read
$\underline{x}_o(t)$ instead of $\underline{x}(t)$,

p.99, chapter 3.6., 6th line from above, read
players instead of playes ,

p.99, 1st line after equation (3.42), read
Further instead of Furhter ,

p.100, 4th line after equation (3.47), read
maximin-value instead of maximum-value .

J.E. Rijnsdorp

List of ERRATA for and supplements to

J.E. Rijnsdorp
MULTIVARAIBLE SYSTEMS IN PROCESS CONTROL

p.136, line 19, new paragraph **from** Before .... ,

p.137, line 1, read

.... "homeostasis" insensitiveness .... ,

p.139, in equation (1) read

$\dot{x}$ = ... instead of x = ... ,

p.139, line 11 from below, new paragraph from More.. ,

p.142, end sentence above formula (8) with

: instead of . ,

p.143, line 20, read

non-minimum instead of non-minium ,

p.144, line 12 from below, read

.. root-locus diagrams (see also Niederliński / 100 /). ,

p.146, line 21, read

... papers about industrial applica- .... ,

p.147, equation (18), read

$$(sI-Q^D_{cc}(s))\,c(s) = Q^D_{cu}(s)\Big[Q^N_{cu}(s)\,u(s)+Iu(s)+Q_{cv}(s)\,v(s)$$

$$+Q_{cm}(s)\,m(s)+Q^N_{cc}(s)\,c(s)+Q_{cw}(s)\,w(s)\Big] ,$$

p.151, line 16, delete final paranthesis .

p.157, figure 8 should be corrected as shown

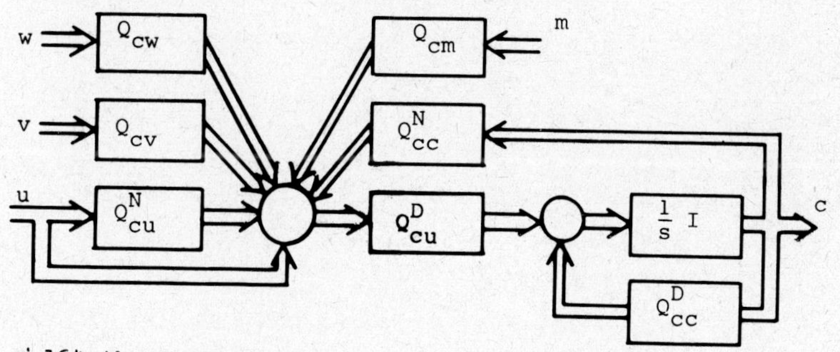

p.164, the following reference should be added:

/ 100 / Niederliński, A.: Regelungstechnik **18** (1970)
No. 12, 551-555

J.P. Waha

List of ERRATA for

J.P. Waha
MULTIVARIABLE TECHNICAL CONTROL SYSTEMS.
SURVEY OF APPLICATIONS IN POWER PLANTS AND POWER DISTRIBUTION SYSTEMS

p.188, line 4, read

... based on Park equations were made. A simulation of 50 outages ... ,

p.188, line 8, read /50/ instead of /72/ ,

p.192, line 18, read /7/ instead of / / ,

p.194, FIG. 10.,

line 3, read SECURITY CONSTRAINTS ,

line 5, read METHOD ,

Caption should be:

LINEAR PROGRAMMING APPROACH OF LOAD SCHEDULING WITH SECURITY CONSTRAINTS .